INDEX TO BOTANICAL MONOGRAPHS

INDEX TO BOTANICAL MONOGRAPHS

A Guide to Monographs and Taxonomic Papers Relating to Phanerogams and Vascular Cryptogams Found Growing Wild in the British Isles

Compiled by
DOUGLAS H. KENT

1967

Published for the
BOTANICAL SOCIETY OF THE BRITISH ISLES
by
ACADEMIC PRESS
LONDON and NEW YORK

ACADEMIC PRESS INC. (LONDON) LTD.
BERKELEY SQUARE HOUSE
BERKELEY SQUARE
LONDON, W.1.

U.S. Edition published by
ACADEMIC PRESS INC.
111 FIFTH AVENUE
NEW YORK, NEW YORK 10003

Library of Congress Catalog Card Number: 67-30764

PRINTED IN GREAT BRITAIN BY
ADLARD & SON, LTD., BARTHOLOMEW PRESS, DORKING, SURREY

Contents

INTRODUCTION vii
ABBREVIATIONS OF THE TITLES OF PERIODICALS, ETC. 1
INDEX TO BOTANICAL MONOGRAPHS 25
INDEX TO FAMILIES AND GENERA, ETC. 155

Introduction

The vast amount of periodical botanical literature which has appeared since the time of Linnaeus contains numerous monographs and taxonomic papers on plant families and genera. So far as I am aware no serious attempt has been made to extract and present in systematic order references to such works as are relevant to the flora of the British Isles since B. D. Jackson's *Guide to the Literature of Botany* published in 1881, and even in this book the data given are often very sketchy.

The present work contains nearly 1,900 references to monographs, taxonomic and cyto-taxonomic papers which have been published since 1800, though a few important monographs which appeared prior to that date have been deemed worthy of inclusion. Such information as is provided, which is acknowledgedly incomplete, was accumulated by searching through runs of over 300 periodical publications in such spare time as I was able to devote to the project between 1957 and 1966.

Although this book is primarily designed for the use of students of the flora of the British Isles, many genera from continental Europe, North America and elsewhere which occur as adventives in our islands are included, and it may, therefore, be found useful to botanists in most temperate regions of the world. Papers which deal exclusively with cytology are omitted, as are, with certain exceptions, monographic accounts of individual species; the exceptions being monotypic genera and collective species (e.g. *Ranunculus auricomus*, *Cardamine pratensis*, *Viola tricolor*, etc.).

Many Floras, etc., contain monographic accounts, but these are mostly well known, and to repeat all the data here would have greatly increased the size and price of this book. I have, however, provided detailed references to H. G. A. Engler's *Das Pflanzenreich*, and have cited also some major floras, as for example G. Hegi's *Illustrierte Flora von Mittel-Europa*. A short list of some of the major works containing monographs is given below:

Ascherson, P. F. A. and Graebner, K. O. P. P. *Synopsis der mitteleuropäischen Flora*. Leipzig. 1896–1938. Vol. 1. Pp. 1–160. 1896 : 161–416. 1897. Vol. 2 (1). Pp. 1–64. 1898 : 65–304. 1899 : 305–544. 1900 : 545–704. 1901 : 705–796. 1902. Vol. 2 (2). Pp. 1–144. 1902 : 145–224. 1903 : 225–530. 1904. Vol. 3. Pp. 1–320. 1905 : 321–560. 1906 : 561–934. 1907. Vol. 4. Pp. 1–80. 1908 : 81–320. 1909 : 321–400. 1910 : 401–640. 1911 : 641–800. 1912 : 801–

886.1913. Vol. 5 (1). Pp. 1–224.1913 : 225–400.1914 : 401–480.1915 : 481–544.1916 : 545–625.1917 : 626–784.1918 : 785–948.1919. Vol. 5 (2). Pp. 1–160.1920 : 161–400.1921 : 401–480.1922 : 481–560.1923 : 561–640. 1926 : 641–812.1929. Vol. 5 (3). Pp. 1–98.1935. Vol. 5 (4). Pp. 1–160. 1936 : 161–252.1938. Vol. 6 (1). Pp. 1–64.1900 : 65–240.1901 : 241–560. 1902 : 561–640.1903 : 641–800.1904 : 801–896.1905. Vol. 6 (2). Pp. 1–160.1906 : 161–496.1907 : 497–688.1908 : 689–1008.1909 : 1009–1094. 1910. Vol. 7. Pp. 1–80.1913 : 81–240.1914 : 241–320.1915 : 321–400. 1916 : 401–480.1917. Vol. 12 (1). Pp. 1–80.1922 : 81–160.1924 : 161–400. 1929 : 401–492.1930. Vol. 12 (2). Pp. 1–160.1930 : 161–480.1931 : 481–640.1934 : 641–790.1935. Vol. 12 (3). Pp. 1–320.1936 : 321–480.1937 : 481–708.1938. Edition 2. Leipzig. 1912–13. Vol. 1. Pp. 1–480.1912 : 481–630.1913.

Clapham, A. R., Tutin, T. G. and Warburg, E. F. *Flora of the British Isles*. Pp. li + 1591. Cambridge, 1952. Edition 2. Pp. xlvii + 1269. 1962.

De Candolle, A. P. *Prodromus Systematis naturalis Regni Vegetabilis, sive Enumeratio contracta Ordinum, Generum, Specierumque Plantarum hucusque cognitorum juxta Methodi naturalis Normas digesta*. Parisiis. 1824–74. Vol. 1. 1824. Vol. 2, 1825. Vol. 3, 1828. Vol. 4, 1830. Vol. 5, 1836. Vol. 6, 1838. Vol. 7 (1), 1838. Vol. 7 (2), 1839. Vol. 8, 1844. Vol. 9, 1845. Vol. 10, 1846. Vol. 11, 1847. Vol. 12, 1848. Vol. 13 (1), 1852. Vol. 13 (2), 1849. Vol. 14. Pp. 1–492.1856 : 493–706.1857. Vol. 15 (1), 1864. Vol. 15 (2). Pp. 1–188. 1862 : 189–1286.1866. Vol. 16 (1), 1869. Vol. 16 (2). Pp. 1–160.1864 : 161–691.1868. Vol. 17, 1873. Index. 1–4. 1843 : 5–7 (1).1840 : 7 (2)–13.1858–59 : 14–17.1874.

Engler, H. G. A. (Ed.). *Das Pflanzenreich. Regni Vegetabilis Conspectus*. Leipzig and Berlin. 1900–53. For details of authors and dates of publication of individual volumes see Davis, M. T. "A Guide and an analysis of Engler's 'Das Pflanzenreich'." Taxon 6, 161–182.1957.

Engler, H. G. A. and Prantl, K. A. E. *Die natürlichen Pflanzenfamilien nebst ihren Gattungen und wichtigeren Arten insbesondere den Nutzpflanzen*. Leipzig. 1887–1915. Vol. 2 (1). Pp. 1–172.1887 : 173–262.1889. Vol. 2 (2). Pp. 1–96.1887 : 97–130.1888. Vol. 2 (3). Pp. 1–144.1887 : 145–168.1889. Vol. 2 (4). Pp. 1–48.1887 : 49–78.1888. Vol. 2 (5). Pp. 1–144.1887 : 145–162. 1888. Vol. 2 (6). Pp. 1–144.1888 : 145–244.1889. Vol. 3 (1). Pp. 1–48. 1887 : 49–144.1888 : 145–289.1889. Vol. 3 (1a). Pp. 1–48.1892 : 49–130. 1893. Vol. 3 (1b). 1889. Vol. 3 (2). Pp. 1–96.1888 : 97–144.1889 : 145–281.1891. Vol. 3 (2a). Pp. 1–48.1890 : 49–142.1891. Vol. 3 (3). Pp. 1–48. 1888 : 49–112.1891. Vol. 3 (4). Pp. 1–94.1890 : 95–362.1896. Vol. 3 (5). Pp. 1–96.1890 : 97–128.1891 : 129–224.1892 : 225–272.1893 : 273–416. 1895 : 417–468.1896. Vol. 3 (6). Pp. 1–96.1890 : 97–240.1893 : 241–340. 1895. Vol. 3 (6a). Pp. 1–96.1893 : 97–254.1894. Vol. 3 (7). Pp. 1–48.1892 : 49–241.1893. Vol. 3 (8). Pp. 1–48.1894 : 49–144.1897 : 145–274.1898.

Vol. 4 (1). Pp. 1–96. 1889 : 97–144. 1890 : 145–183. 1891. Vol. 4 (2). Pp. 1–48. 1892 : 49–310. 1895. Vol. 4 (3a). Pp. 1–48. 1891 : 49–96. 1893 : 97–224. 1895 : 225–320. 1896 : 321–384. 1897. Vol. 4 (3b). Pp. 1–96. 1891 : 97–144. 1893 : 145–240. 1894 : 241–378. 1895. Vol. 4 (4). 1891. Vol. 4 (5). Pp. 1–80. 1889 : 81–272. 1890 : 273–304. 1892 : 305–368. 1893 : 369–402. 1894. *Gesamtregister*. Pp. 1–320. 1898 : 321–462. 1899. *Nachtrage*. Vol. 1. 1897. Vol. 2. 1900. Vol. 3. Pp. 1–192. 1906 : 193–288. 1907 : 289–379. 1908. Vol. 4. Pp. 1–192. 1914 : 193–381. 1915. Edition 2. Leipzig and Berlin. 1925→. Vols. 13, 14a, 14e, 15a, 16b, 16c, 17b, 18a, 19a, 19b1, 19c, 20b, 21. Leipzig. 1926. Vols. 14d, 17aII, 20d. Berlin. 1926. Vols. 13, 14a, 1926. Vol. 14d. 1956. Vol. 14e. 1940. Vol. 15a. 1930. Vol. 16b. 1935. Vol. 16c. 1934. Vol. 17aII. 1959. Vol. 17b. 1936. Vol. 18a. 1930. Vol. 19a. 1931. Vol. 19bI. 1940. Vol. 19c. 1931. Vol. 20b. 1942. Vol. 20d. 1953. Vol. 21. 1925.

HEGI, G. *Illustrierte Flora von Mittel-Europa*. München. 1906–31. Vol. 1. Pp. 1–72. 1906 : 73–312. 1907 : 313–412. 1908. Vol. 2. Pp. 1–128. 1908 : 129–408. 1909. Vol. 3. Pp. 1–36. 1909 : 37–328. 1910 : 329–472. 1911 : 473–608. 1912. Vol. 4 (1). Pp. 1–96. 1913 : 97–144. 1914 : 145–192. 1916 : 193–320. 1918 : 321–491. 1919. Vol. 4 (2). Pp. 497–540. 1921 : 541–908. 1922 : 909–1112b. 1923. Vol. 4 (3). Pp. 1113–1436. 1923 : 1437–1748. 1924. Vol. 5 (1). Pp. 1–316. 1924 : 317–674. 1925. Vol. 5 (2). Pp. 679–994. 1925 : 995–1562. 1926. Vol. 5 (3). Pp. 1567–1722. 1926 : 1723–2250. 1927. Vol. 5 (4). Pp. 2255–2632. 1927. Vol. 6 (1). Pp. 1–112. 1913 : 113–304. 1914 : 305–352. 1915 : 353–400. 1916 : 401–496. 1917 : 497–544. 1918. Vol. 6 (2). Pp. 549–1152. 1928 : 1153–1386. 1929. Vol. 7. 1931. Edition 2. München. 1936→. Vol. 1. 1936. Vol. 2. 1939. Vol. 3 (1). Pp. 1–240. 1957 : 241–452. 1958. Vol. 3 (2). Pp. 453–532. 1959 : 533–692. 1960 : 693–772. 1961 : 773–852. 1962. Vol. 3 (3). 1965. Vol. 4 (1). Pp. 1–80. 1958 : 81–160. 1959 : 161–320. 1960 : 321–400. 1961 : 401–480. 1962 : 481–567. 1963. Vol. 4 (2). Pp. 1–80. 1961 : 81–224. 1963. Vol. 4 (3). 1964. Vol. 5. 1964–66. Vol. 6. 1964–66.

TUTIN, T. G., HEYWOOD, V. H., BURGESS, N. A., VALENTINE, D. H., WALTERS, S. M., WEBB, D. A. *et al.* (Eds.). *Flora Europaea*. Cambridge. Vol. 1. Lycopodiaceae to Platanaceae Pp. xxxii + 464. 1964.

As a simple guide much useful information may also be obtained from books under the following classified headings.

Trees and Shrubs

BEAN, W. J. *Trees and Shrubs Hardy in the British Isles*. Vol. 1. Pp. vii + 688. Vol. 2. Pp. 736. London. 1914. Vol. 3. Pp. xiv + 517. 1933. Edition 2. 2 vols. 1916. Edition 3. 2 vols. 1921. Edition 4. 2 vols. 1925. Edition 5. 2 vols. 1929. Edition 6. 3 vols. 1933. Edition 7. 3 vols. 1950–51.

BOOM, B. K. *Flora der Cultuurgewassen von Nederland*. Vol. 1. *Nederlandse Dendrologie*. Edition 3. Pp. 444. Wageningen. 1949. Edition 5. Pp. 465. 1965.

ELWES, H. J. and HENRY, A. *The Trees of Great Britain and Ireland.* 7 vols. + index. Pp. 2022 + 412 plates. Edinburgh. 1906–13.

GILBERT-CARTER, H. *Our Catkin-bearing Plants.* Pp. xii + 61. Oxford. 1930. Edition 2. Pp. xii + 61. London. 1932.

GILBERT-CARTER, H. *British Trees and Shrubs, including those commonly planted: a Systematic Introduction to our Conifers and Woody Dicotyledones.* Pp. xv + 291. London. 1936.

KRÜSSMANN, G. *Handbuch der Laubgehölze.* Vol. 1. Pp. vi + 495. Berlin and Hamburg. 1960. Vol. 2. Pp. 608. 1962.

MAKINS, F. K. *The Identification of Trees and Shrubs.* Pp. 326. London. 1936. Reprinted with corrections. Pp. vii + 326. 1944. Edition 2. Pp. vii + 375. 1948.

REHDER, A. *A Manual of Cultivated Trees and Shrubs Hardy in North America.* Pp. xxxvii + 930. Washington. 1927. Edition 2. Pp. xxx + 996. 1940.

SCHNEIDER, K. C. *Illustriertes Handbuch der Laubholzkunde.* Vol. Pp. iv + 1–448. Jena. 1904 : 449–592. 1905 : 593–810. 1906. Vol. 2. Pp. 1–240. 1907 : 241–496. 1909 : 497–816. 1911 : 817–1070. 1912. Register. Pp. 136. 1912.

Cultivated Plants

BAILEY, L. H. *Manual of Cultivated Plants.* Pp. 851. New York. 1924. Edition 2. Pp. 1116. 1949.

BOOM, B. K. *Flora der Cultuurgewassen von Nederland.* Vol. 2. *Flora der Gekweekte Kruidachtige gewassen.* Pp. 450. Wageningen. 1950.

CHITTENDEN, F. J. (Ed.). *The Dictionary of Gardening. A Practical and Scientific Encyclopaedia of Horticulture.* Royal Horticultural Society. 4 vols. Pp. xvi + 2316. Oxford. 1951.

SYNGE, P. M. (Ed.). *Supplement to the Dictionary of Gardening. A Practical and Scientific Encyclopaedia of Horticulture.* Royal Horticultural Society. Pp. vii + 334. Oxford. 1956.

Aquatic Plants

FASSETT, N. C. *Manual of Aquatic Plants.* Pp. 382. New York. 1940.

GLÜCK, H. *Pteridophyten und Phanerogamen,* in PASCHER, A. (Ed.). *Die Süsswasser-flora Mitteleuropas* 15. Pp. xx + 486. Jena. 1936.

The list of genera used in this book is arranged in the order of J. E. Dandy's *List of British Vascular Plants* (1958), and the nomenclature, though based on that work, has been brought up to date. Where, however, a generic name has been altered or amended, the name used in the *List of British Vascular Plants* is given in brackets. It must, however, be pointed out that the changes in the names of certain genera include authentic nomenclatural changes and changes based on taxonomic opinion, the latter of which may or may not be acceptable to the reader. In addition a number of genera are included which do not appear in the *List of British Vascular Plants*; these contain species which occur as established or casual adventives in the British Isles.

Most of the books and papers cited have been seen by the compiler, though in the latter instance sometimes only as reprints from obscure periodicals. A number of titles gleaned from the literature have not been seen, and these are prefixed by an asterisk.

Titles of books and periodicals given in a cyrillic alphabet have been transliterated, as have works in Chinese and Japanese.

Details of monographs and taxonomic papers not given in the following pages would be gratefully received by the compiler.

I am deeply indebted to the many friends and correspondents who have kindly provided suggestions as to periodicals worth searching, sources of reference, and individual references, etc., in particular Dr. R. K. Brummitt, Dr. C. D. K. Cook, A. C. Jermy, J. E. Lousley, the late N. Y. Sandwith, Dr. W. T. Stearn and the late Dr. E. F. Warburg. Grateful thanks are also due to J. E. Dandy for advice on nomenclature, and to J. C. Gardiner for his encouragement and kindness in assisting in the publication of this book.

75 *Adelaide Road,*
West Ealing,
*London, W.*13
August 1967

DOUGLAS H. KENT

Abbreviations of the Titles of Periodicals

Abh. Böhm. Ges. Wiss. (Math.-Nat.)
Abhandlungen der Königlich-böhmischen Gesellschaft der Wissenschaften (Math.-Nat. Kl.). Praha. 1775–1891, 1928–

Abh. K. Leop.-Carol. Deutsch. Akad. Nat.
Abhandlungen der Kaiser Leopoldinisch-Carolinischen Deutschen Akademie der Naturforscher. Halle. 1757–1928. Continued as *Abhandlungen der Deutschen Akademie der Naturforscher Leopoldina*. Halle. 1932–

Abh. Nat. Ver. Bremen
Abhandlungen hrsg. vom Naturwissenschaftlichen Verein zu Bremen. Bremen. 1868–

Abh. Nat. Ver. Hamburg
Abhandlungen aus dem Gebiete der Naturwissenschaftlicher Verein von Hamburg. Hamburg. 1846–82.

Abh. Preuss. Akad. Wiss.
Abhandlungen der Preussischen Akademie der Wissenschaften. Phys.-math. Kl. Berlin. 1804–1944. Continued as *Abhandlungen der Deutschen Akademie der Wissenschaften zu Berlin*. Berlin. 1945–

Abh. Sächs. K. Ges. Wiss.
Abhandlungen der mathematisch-physischen Classe der Sächsischen Königlich Gesellschaft der Wissenschaften. Leipzig. 1852–

Abh. Vortr. Ges. Naturw. (Berlin)
Abhandlungen und Vorträge zur Geschichte der Naturwissenschaften. Berlin.

Abh. Zool.-Bot. Ges. Wien
Abhandlungen der Kaiserlich-königliche Zoologisch-botanischen Gesellschaft in Wien. Wien. 1901–

Act. Soc. Linn. Bordeaux
Actes de la Société Linnéenne de Bordeaux. Bordeaux. 1830–

Acta Acad. Sci. Nat. Morav.-Siles.
Acta Societatis scientiarum naturalium moravo-silesiacae. Brno. 1924–47. Continued as *Acta Academiae scientiarum naturalium moravo-silesiacae*. Brno. 1948–53. Continued as *Práce Brněnské Základny Československé akademie věd*. Brno. 1954–

Acta Biol. Cracov.
Acta biologica cracoviensia. Kraków. 1958–

Acta Bot. Bohemica
Acta botanica Bohemica. Praha. 1922–47.

Acta Bot. Fenn.
Acta botanici Fennica. Helsingforsiae. 1925–

Acta Bot. Hung.
Acta botanica Academiae Scientiarum Hungaricae. Budapest. 1955–

Acta Bot. Neerl.
Acta botanica Neerlandica. Amsterdam. 1952–

Acta Bot. Zagreb.
Acta botanica Instituti botanici Universitatis Zagrebiensis. Zagreb. 1925–56. Continued as *Acta botanica Croatica*. Zagreb. 1957–

Acta Dendr. Čechoslov.
Acta dendrologica Čechoslovaca/Dendrologický Sbornik. Opava. 1958–

Acta Flora Suec.
Acta Flora Suecica. Vol. 1. Stockholm. 1921.

Acta Geobot. Hung.
Acta geobotanica Hungarica. Debrecen. 1936–49.

Acta Hort. Berg.
Acta Horti Bergiani. Stockholm. 1890–

Acta Hort. Bot. Pragensis
Acta Horti botanici Pragensis. Praga. 1962–

Acta Hort. Bot. Univ. Latv.
Acta Horti botanici Universitatis Latviensis. Riga. 1926–44.

Acta Hort. Petrop.
Acta Horti Petropolitani. Peterburgi. 1871–1918. Continued as *Trudy glavnago botaničeskago Sada*. Petrograd. 1918–31.

Acta Inst. Bot. Acad. Sci. URSS.
Acta Instituti botanici Academiae Scientiarum URSS. Leningrad and Mosqva. 1933–

Acta Phyt. Suec.
Acta Phytogeographica Suecica. Uppsala. 1921–

Acta Rer. Nat. Distr. Ostrav.
Acta rerum naturalium Districtus ostraviensis/Sbornik Přirodovědecké společnosti v Mor. Ostravé. Mor. Ostrava. 1921–46. Continued as *Přirodovědecký Sbornik Ostravského kraje*. Opava. 1949–58. Continued as *Přirodovedný Časopis Slezský*. Opava. 1959–

Acta Soc. Bot. Fenn.
Acta Societatis Botanica Fennica. Helsinki. 1844–

Acta Soc. Bot. Pol.
Acta Societatis Botanicorum Poloniae. Warszawa. 1923–

Acta Soc. Fauna Fl. Fenn.
Acta Societatis pro Fauna et Flora Fennica. Helsingforsiae. 1875–

Acta Soc. Sci. Fenn.
Acta Societatis Scientiarum Fennicae. Helsingforsiae. 1842–1926.

Acta Univ. Carolinae
Acta Universitatis Carolinae. Biologica. Prague. 1954–

Agr. Trop.
Agronomie Tropicale. Nogent-sur-Marne. 1946–

Agrártud. Egyetem Kert Kar. Közl.
Agrártudományi egyetem Kert-és Szölögazdaságtudományi Karának Közlemenyei. Budapest. 1948–52. Continued as *Kertészeti és Szölészeti föiskola évkonyve*. Budapest. 1953–

Agron. Lusit.
Agronomia Lusitana. Savacém. 1939–

Agros
Agros. Lisboa. 1917–

Aliso
El Aliso. Anaheim, Claremont. 1948–. *Aliso* from vol. 5 (1961).

Allgem. Bot. Zeitschr.
Allgemeine botanische Zeitschrift für Systematik, Floristik, Pflanzen-geographie. Karlsruhe. 1895–1927.

Amer. Fern J.
American Fern Journal. Port Richmond, New York. 1910–

Amer. J. Bot.
American Journal of Botany. Lancaster, Pennsylvania. 1914–

Amer. Midl. Nat.
American Midland Naturalist. Notre Dame, Indiana. 1909–

An. Acad. Rep. Pop. Române
Analele Academiei Republici Populare Române, seria Geologie, Geografie, Biologie, Stiinte technice si Agricole. Bucureşti. 1948–

Anais Inst. Sup. Agron. (Lisboa)
Anais do Instituto superior de Agronomia. Lisboa. 1920–

Anal. Acad. Române
Analele Academiei Române. Bucureşti. 1867–

Anal. Inst. Biol. (Mexico)
Anales del Instituto de Biologica. Mexico. 1930–

Anal. Inst. Bot. Cav.
Anales del Jardín botánico de Madrid. Vols 1–9. Madrid. 1941–50. Continued as *Anales del Instituto botánico A. J. Cavanilles*. Vol. 10– Madrid. 1950–

Ann. Bot.
Annals of Botany. London. 1887–

Ann. Bot. (London)
Annals of Botany. London. Vols 1–2. London. 1805–6.

Ann. Bot. (Roma)
Annali di Botanico. Roma. 1903–

Ann. Bot. Fenn.
Annales botanici Fennici. Helsinki. 1964–

Ann. Bot. Soc. Zool.-Bot. Fenn. Vanamo
Annales botanici Societatis zoologicae-botanicae Fennicae "Vanamo". Helsinki. 1923–63. Continued as *Annales botanici Fennici*. Helsinki. 1964–

Ann. Conserv. Jard. Bot. Genève
Annuaire du Conservatoire et du Jardin botanique de Genève. Genève. 1897–1922. Continued as *Candollea*. Genève. 1922–

Ann. Inst. Nat. Rech. Agron.
Annales de l'Institut national de la recherche agronomique. Paris. 1950–

Ann. Mag. Nat. Hist.
Annals and Magazine of Natural History. London. 1841–67.

Ann. Missouri Bot. Gard.
Annals of the Missouri Botanical Garden. St. Louis, Missouri. 1914–

Ann. Mus. Hung.
Annales historico-naturales Musei nationalis Hungarici. Budapest. 1903–

Ann. Nat. Mus. (Wien)
Annalen des K.K. naturhistorischen (Hof) museums. Wien. 1886–

Ann. New York Acad. Sci.
Annals of the New York Academy of Sciences. New York. 1823–

Ann. Rep. Proc. Bristol Nat. Soc.
Annual Report and Proceedings of the Bristol Naturalists' Society. Bristol. 1863–

Ann. Rep. Missouri Bot. Gard.
Annual Report of the Missouri Botanical Garden. St. Louis. 1890–1912. Continued as *Annals of the Missouri Botanical Garden*. St. Louis. 1914–

Ann. Rep. Smith. Inst.
Annual Report of the Board of Regents of the Smithsonian Institution. Washington. 1846–

Ann. Sci. Nat.
Annales des sciences naturelles. Paris. 1824–

Ann. Univ. Debrec.
Annales biologicae Universitatis debreceniensis olim Tisia resp. Acta geobotanica hungarica. Budapest. 1950–

Ann. Univ. Mariae-Curie
Annales Universitatis Mariae Curie-Sklodowska. Lublin. 1946–

Ann. Univ. Sci. Budapest.
Annales Universitatis scientiarum budapestinensis de Rolando Eötvös nominatae: sectio biologica. Budapest. 1957–

Arbeit. Bot. Gart. Univ. Breslau
Arbeiten aus den königl. Botanischen Garten zu Breslau. Breslau. 1892–?

Arch. Bot.
Archivio botanico par la sistematica, fitogeografia e genetica. Forli. 1925–

Arch. Freunde Nat. Mecklenburg
Archiv der Freunde der Naturgeschichte in Mecklenburg. Rostock. 1954–

Arch. Soc. Zool.-Bot. Fenn. Vanamo
Archivum Societatis Zoologico-botanicae Fennicae "Vanamo". Helsinki. 1946–

Arkiv. Bot.
Arkiv för Botanik. Uppsala. 1903–

Atti Accad. Sci. Veneto-Trentino-Istriana
Atti dell'Accademia scientifica veneto-trentino-istriana. Padova. 1872–1935.

Atti Ist. Bot. Pavia
Atti dell'Istituto botanico "Giovanni Briosi" e Laboratoria crittogamico della Università di Pavia. Milano. 1888–

Atti Reale Accad. (Palermo)
Atti della Reale Accademia di Scienze, Lettere e Belle Arti. Palermo. 1845–

Austral. J. Bot.
Australian Journal of Botany. Melbourne. 1953–

Baileya
Baileya. Quarterly journal of horticultural taxonomy. New York. 1953–

Bauhinia
Bauhinia. Zeitschrift der Basler botanischen Gesellschaft. Basel. 1955–

Beih. Bot. Centr.
Beihefte zum botanischen Centralblatt. Cassel, Jena and Dresden. 1891–1944.

Beih. Nova Hedw.
Beihefte Nova Hedwigia. Weinheim. 1962–

Beitr. Biol. Pflanzen
Beiträge zur Biologie der Pflanzen. Breslau. 1870–

Beitr. Bot. (Leipzig)
Beiträge zur Botanik. Leipzig. 1842.

Ber. Bayer. Bot. Ges.
Berichte der bayerischen botanischen Gesellschaft zur Erforschung der heimischen Flora. München. 1891–

Ber. Deutsch. Bot. Ges.
Berichte der deutschen botanischen Gesellschaft. Berlin. 1883–

Ber. Schweiz. Bot. Ges.
Berichte der schweizerischen botanischen Gesellschaft. Bern. 1891–

Bergens Mus. Aarb.
Bergens Museum Aarbog. Bergen. 1886–

Bibl. Bot.
Bibliotheca botanica. Abhandlungen aus dem Gesammtgebiete der Botanik. Stuttgart. 1886–

Biol. Jaarb.
Botanisch Jaarboek. Gand. 1899–1934. Continued as *Biologisch Jaarboek*. Antwerpen. 1934–

Biol. Medd.
Biologiske Meddelelser. Kongelige Danske Videnskabernes Selskab. Kjøbenhavn. 1917–

Biol. Skr. Danske Vid. Selsk.
Biologiske Skrifter K. Danske Videnskabernes Selskab. København. 1933–

Biol. Zentralbl.
Biologisches Zentralblatt. Leipzig. 1881–

Blätt. Staudenk.
Blätter für Staudenkunde. Berlin. 1937–

Blumea
Blumea. Tijdschrift voor de Systematiek en de Geografie der Planten. Leiden. 1934–

Boissiera
Boissiera. Supplement of *Candollea*. Genève. 1936–

Bol. Inst. Forest. Invest. Exp.
Boletín Instituto Forestal de Investigaciones y Experiencias. Madrid. 1928–

Bol. Inst. Nac. Invest. Agron.
Boletín del Instituto Nacional de Investigaciones Agronómicas. Madrid. 1935–

Bol. Soc. Brot.
Boletim da Sociedade Broteriana. Coimbra. 1880–

Boll. Orto Bot. Palermo
Bollettino de R. Orto botanico e Giardino coloniale di Palermo. Palermo. 1897–1921.

Boll. Soc. Bot. Ital.
Bollettino della Società botanica Italiana. Firenze. 1892–1926.

Borbasia
Borbasia. Dissertationes botanicae. Budapest. 1938–

Bot. Arch.
Botanisches Archiv. Berlin. 1922–

Bot. Archiv.
Botanisches Archiv. Zeitschrift für die Botanik. Königsberg.

Bot. Centralbl.
Botanisches C(Z)entralblatt. Jena and Dresden. 1880–

Bot. Centralbl. Beih.
Beihefte zum Botanischen Centralblatt. Cassel. 1891–1943.

Bot. Gaz.
Botanical Gazette. Chicago. 1875–

Bot. Jahrb.
Botanische Jahrbücher für Systematik, Pflanzengeschichte und Pflanzengeographie. Leipzig. 1880–

Bot. Közl.
Botanikai Közlemények. Budapest. 1902–

Bot. Mag.
Botanical Magazine, or Curtis's Botanical Magazine. London. 1793–

Bot. Mag. (Tokyo)
Botanical Magazine. Tokyo. 1887–

Bot. Not.
Botaniska Notiser. Lund. 1839–

Bot. Tidsskr.
Botanisk Tidsskrift. Kjøbenhavn. 1866–

Bot. Zeit.
Botanische Zeitung. Berlin and Leipzig. 1843–1910.

Bot. Žurn.
Journal de la Société botanique de Russie. Vols 1–16. Petrograd. 1916–31. Continued as *Botaničeskij Žurnal SSSR.* Vols 17–32. Leningrad and Moskva. 1932–47. Continued as *Botaničeskij Žurnal Akad. Nauk SSSR.* Vol. 33–. Leningrad. 1948–

Bothalia
Bothalia. National Herbarium, Pretoria. Pretoria. 1921–

Brit. Fern Gaz.
British Fern Gazette. Kendal. 1909–

Brittonia
Brittonia. New York. 1931–

Broteria
Broteria. Lisboa. 1902–

Bul. Acad. Inalt. Stud. Agron. Cluj
Buletinul Academiei de Inalte Studii Agronomice din Cluj. Cluj. 1930–37.

Bul. Fac. Sti. Cernăuti
Buletinul Facultăţii de ştiinţe din Cernăuti. Cernăuti. 1923–37.

Bul. Grăd. Bot. Cluj
Buletinul de informaţii al Grădinii botanice şi al Muzeului botanic de la Universitatea din Cluj. Vols 1–5. Cluj. 1921–25. Continued as *Buletinul Grădinii botanice şi al Muzeului botanic de la Universitatea din Cluj.* Vol. 6–. Cluj. 1926–

Bull. Acad. Polon. Sci. Lett.

Bulletin international de l'Académie polonaise des sciences et des lettres. Cracovie. 1919–51. Continued as *Bulletin de l'Académie polonaise des sciences.* Varsovie. 1953–

Bull. Acad. Sci. Cracov.

Bulletin international Akademija Umiejetnósci . . . Comptes rendus des Séances. Cracovie. 1889–1918.

Bull. Ass. Russe Rech. Sci. (Prague)

Bulletin de l'Association russe pour les recherches scientifiques à Prague/Zapiski. Nauchno-issledovatel'skoe ob''edinenie, Russkiĭ svobodnȳĭ universitet v Prage. Praha. 1935–

Bull. Brit. Mus. (Bot.)

Bulletin of the British Museum (Natural History). Botany. London. 1951–

Bull. Centre Rech. Sci. Biarritz

Bulletin du Centre d'études et des recherches scientifiques. Biarritz. 1956–

Bull. Forest Exp. Stat. Meguro, Tokyo

Bulletin. Forest Experimental Station, Meguro, Tokyo. Tokyo.

Bull. Herb. Boiss.

Bulletin de l'Herbier Boissier. Genève and Bâle. 1893–1908.

Bull. Inst. Bot. (Sofia)

Bulletin de l'Institut botanique. Sofia. 1950–

Bull. Iraq Mus.

Bulletin of the Iraq Natural History Museum. University of Baghdad. Baghdad. 1961–

Bull. Jard. Bot. Bruxelles

Bulletin du Jardin botanique de l'État à Bruxelles. Bruxelles. 1902–

Bull. Jard. Bot. URSS.

Bulletin du Jardin botanique de l'Académie des Sciences de l'URSS/Bulletin du Jardin botanique de la République russe (de l'URSS). Petrograd. 1918–32.

Bull. Mus. Hist. Nat. (Paris)

Bulletin du Muséum, national d'Histoire naturelle. Paris. 1895–

Bull. New York State Coll. Forestry

Bulletin of the New York State College of Forestry. Syracuse, New York. 1913–

Bull. New York State Exper. Station

Bulletin of the New York State Agricultural Experiment Station. Geneva. 1885–

Bull. Sci. Acad. Imp. Sci. St. Pétersb.

Bulletin scientifique l'Académie impériale des sciences de Saint-Pétersbourg. Saint Pétersbourg and Leipzig. 1835–?

Bull. Soc. Agric. Sci. Lit. Pyrénées-Orient.

Bulletin de la Société agricole, scientifique et littéraire des Pyrénées-orientales. Perpignan. 1835–1912.

Bull. Soc. Bot. Belg.
Bulletin de la Société Royal de botanique de Belgique. Bruxelles. 1862–

Bull. Soc. Bot. Bulg.
Bulletin de la Société botanique de Bulgarie. Sofia. 1926–43.

Bull. Soc. Bot. France
Bulletin de la Société botanique de France. Paris. 1854–

Bull. Soc. Bot. Genève
Bulletin de la Société botanique de Genève. Genève. 1879–

Bull. Soc. Dendr. France
Bulletin de la Société dendrologique de France. Paris. 1906–

Bull. Soc. Hist. Nat. Afr. Nord
Bulletin de la Société d'Histoire naturelle de l'Afrique du Nord. Alger. 1909–

Bull. Soc. Nat. Luxemb.
Bulletin mensuel de la Société de naturalistes luxemburgeois. Luxemburg. 1907–48. Continued as *Bulletin de la Société des naturalistes luxembourgeois.* Luxemburg. 1949–

Bull. Soc. Nat. Moscou
Bulletin de la Société Impériale des Naturalistes de Moscou: section biologique. Moscou. 1829–

Bull. Soc. Neuchât. Sci. Nat.
Bulletin de la Société Neuchâteloise des sciences naturelles. Neuchâtel. 1844–

Bull. Soc. Phys. Genève
Bulletin de la Société de Physique et d'Histoire Naturelle de Genève. Genève.

Bull. Soc. Sci. Anjou
Bulletin de la Société d'études scientifiques Anjou. Anjou.

Bull. Soc. Sci. Bretagne
Bulletin de la Société scientifique de Bretagne. Rennes. 1924–

Bull. Soc. Sci. Nat. Maroc
Bulletin de la Société des sciences naturelles du Maroc. Rabat. 1921–52. Continued as *Bulletin de la Société des sciences naturelles et physiques du Maroc.* Rabat. 1953–

Bull. Torrey Bot. Club
Bulletin of the Torrey Botanical Club. New York. 1870–

Caldasia
Caldasia. Boletín del Instituto de Ciencias naturales. Universidad nacional de Colombia. Bogota. 1940–

Canad. Field-Nat.
Canadian Field-Naturalist. Ottawa. 1919–

Canad. J. Bot.
Canadian Journal of Botany. Ottawa. 1951–

Canad. J. Gen. Cyt.
Canadian Journal of Genetics and Cytology. Ottawa. 1959–

Candollea
Candollea. Organe du Conservatoire et du Jardin botaniques de la Ville de Genève. Genève. 1922–

Carnegie Inst. Washington Publ.
Carnegie Institution of Washington Publications. Washington. 1902–

Čas. Slez. Mus. Opavaě
Časopis Slezského Musea Opavaě, series A. Praha. 1951–

Chron. Bot.
Chronica Botanica. Leiden, Waltham, Mass. 1935–

Collect. Bot.
Collectanea botanica a Barcinonensi Botanico Instituto edita. Barcinone. 1946–

Comm. Bur. Past. Field Crops Mim. Publ.
Commonwealth Bureau of Pastures and Field Crops. Mimeographed Publications. Hurley, Maidenhead. 1959–

Comm. Fac. Sci. Univ. Ankara (Sci. Nat.)
Communications de la Faculté des Sciences de l'Université d'Ankara. Série C: Sciences naturelles. Ankara. 1948–

Compt. Rend. Trav. Lab. Carlsb. (Sér. Physiol.)
Comptes rendus des Travaux du Laboratoire Carlsberg. Série Physiologique. Vol. 21–. Copenhagen. 1934–. The earlier volumes not in series are sometimes titled *Meddelelser frå Carlsberg Laboratoriet*.

Contr. Bot. (Cluj)
Contribuții Botanice Universitatea "Babes-Bolyai" din Cluj, Grădina Botanică. Cluj.

Contr. Gray Herb.
Contributions from the Gray Herbarium of Harvard University. Cambridge, Mass. 1891–

Contr. Inst. Bot. Univ. Montréal
Contributions de l'Institut botanique de l'Université de Montréal. Montréal. 1938–

Contr. Inst. Hort. Econ. Bot. Taihoku Univ.
Contributions from the Horticultural and Economic Botany Institute, Taihoku Imperial University. Taiwan. 1929–

Contr. Lab. Bot. Univ. Montréal
Contributions du Laboraitoire de botanique de l'Université de Montréal. Montréal. 1922–38. Continued as *Contributions de l'Institut botaniqué de l'Université de Montréal*. Montréal. 1938–

Contr. New South Wales Nat. Herb.
Contributions from the New South Wales National Herbarium. Sydney. 1939–

Contr. U.S. Nat. Herb.
Contributions from the United States National Herbarium. Washington. 1890–

Dansk Bot. Arkiv
Dansk botanisk Arkiv. København. 1913–

Danske Vid. Selsk.
Kongelige Danske Videnskabernes Selskabs Skrifter. København. 1745–1938.

Darwiniana
Darwiniana. Revista del Instituto de botánico Darwinion. Buenos Aires, etc. 1922–

Delpinoa
Delpinoa. Bolletino dell'Orto botanico di Napoli. Napoli. 1948–

Denkschr. Akad. München
Denkschriften der K.B. Akademie der Wissenschaften besonders abgedruckt. München. 1809–25.

Denkschr. Akad. München Abhand.
Denkschriften der K.B. Akademie der Wissenschaften besonders abgedruckt Abhandlung. München. 1831–?

Denkschr. Akad. Wiss. Math.-Nat. Kl. (Wien)
Denkschriften der Kaiserlichen Akademie der Wissenschaften, math.-naturwiss. Kl. Wien. 1850–1946.

Denkschr. Schweiz. Naturf. Ges.
Denkschriften der Schweizerischen naturforschenden Gesellschaft. Zürich. 1920–

Értek. Term. Köréb. Magyar Tud. Akad.
Értekezések a Természettudományok Köréböl Magyar Tudományos Akadémia. Vols 1–23. Budapest. 1867–95.

Féd. Soc. Hort. Belg. Bull.
Bulletin. Fédération des Sociétés d'Horticulture de Belgique. Gand. 1862–88.

Fedde. Rep.
Repertorium Specierum novarum Regni vegetabilis. Vols 1–51. Berlin. 1905–42. Continued as *Feddes Repertortium Specierum novarum Regni vegetabilis.* Vols 52–. Berlin. 1943–

Fedde. Rep. Beih.
Repertorium Specierum novarum Regni vegetabilis. Beiheft. Vols 1–127. Berlin. 1914–42. Continued as *Fedde's Repertorium Specierum novarum Regni vegetabilis. Beiheft.* Vols 128–. Berlin. 1943–

Flora
Flora oder allgemeine botanische Zeitung. Jena and Regensburg. 1818–

Flora Fennica
Flora Fennica. Helsinki. 1923–

Fragm. Fl. Geobot.
Fragmenta floristica et geobotanica. Kraków. 1945–

Gard. Chron.
Gardeners' Chronicle and Agricultural Gazette. London. 1841–

Gartenflora
Gartenflora. Berlin and Erlangen. 1852–1938.

Gentes Herb.
Gentes Herbarum. Occasional Papers on the Kinds of Plants. Ithaca, New York. 1920–

Giorn. Bot. Ital.
Giornale botanico italiano. Firenze. 1844–52.

Gorteria
Gorteria. Tijdschrift ten dienste van de floristiek, de oecologie en het vegetatie-onderzoek van Nederland, uitgegeven door het Rijksherbarium, Leiden. Leiden. 1961–

Heather Soc. Year Book
Heather Society Year Book. London. 1961–

Hedw.
Hedwigia. Organ für Kryptogamenkunde und Phytopathologie. Dresden. 1852–1944. Continued as *Nova Hedwigia. Zeitschrift für Kryptogamenkunde.* Weinheim. 1959–

Helios
Helios. Abhandlungen und Mitteilungen aus der Gesamtgebiet der Naturwissenschaften. Berlin. 1883–1916, 1926–

Herb. Rev.
Herbage Reviews. Aberystwyth. 1933–40.

Herbertia
Herbertia. La Jolla, Stanford. 1936–

Hercynia
Hercynia. Halle. 1936–42.

Hereditas
Hereditas, genetiskt arkiv. Lund. 1920–

Hilgardia
Hilgardia. A Journal of Agricultural Science. California Agricultural Experiment Station. Berkeley. 1925–

Illinois Biol. Monogr.
Illinois Biological Monographs. University of Illinois. Urbana. 1914–

Israel J. Bot.
Israel Journal of Botany. Jerusalem. 1951–

Izvest. Bot. Inst. (Sofia)
Izvestija na Botaničeskija Institut. Bălgarska Akademija na Naukitĕ, etc. Sofija. 1950–

Izvest. Glav. Bot. Sada
Izvestiya Glavnogo Botanitcheskogo Sada SSSR (afterwards *RSFSR*). Sada. 1918–32. Continued as *Sovetskaya Botanika*. Leningrad. 1933–

J. Agric. W. Australia
Journal of the Department of Agriculture for Western Australia. Perth. 1900–51. Continued as *Journal of Agriculture of Western Australia*. 1952–

J. Arnold Arb.
Journal of the Arnold Arboretum, Harvard University. Lancaster, Pennsylvania. 1919–

J. Bot.
Journal of Botany, British and Foreign. London. 1863–1942.

J. Bot. (Paris)
Journal de Botanique. Paris. 1887–1913.

J. Ecol.
Journal of Ecology. Journal of the British Ecological Society. Cambridge and Oxford. 1913–

J. Elisha Mitchell Sci. Soc.
Journal of the Elisha Mitchell Scientific Society. Chapel Hill, North Carolina. 1853–

J. Fac. Agric. Hokkaido Univ.
Journal of the Faculty of Agriculture, Hokkaido (Imperial) University. Sapporo. 1927–

J. Fac. Sci. Tokyo Univ. (Bot.)
Journal of the Faculty of Science, Tokyo Imperial University, sect. 3 *Botany*. Tokyo. 1925–

J. Fac. Text. Techn. Shinshu Univ.
Journal of the Faculty of Textiles and Sericulture, Shinshu University. Ueda. 1951–60. Continued as *Journal of the Faculty of Textile Science and Technology, Shinshu University*. 1961–

J. Gen.
Journal of Genetics. Cambridge and Calcutta. 1910–

J. Jap. Bot.
Journal of Japanese Botany. Tokyo. 1916–

J. Linn. Soc. Bot.
Journal of the Linnean Society of London. Botany. London. 1855–

J. Roy. Hort. Soc.
Journal of the Royal Horticultural Society. London. 1865–

J. S. Afr. Bot.
Journal of South African Botany. Cape Town. 1935–

J. Washington Acad. Sci.
Journal of the Washington Academy of Sciences. Washington. 1911–

Jaarb. Ned. Dendr. Ver.
Jaarboek Nederlandsche Dendrologische Vereeniging. Wageningen. 1925–
Jahrb. Berlin. Bot. Gart.
Jahrbuch des Gartens und des Botanischen Museums K. Friedrich-Wilhelms-Universitaet. Berlin.
Jahrb. Land. Forstw. Fak. Univ. Sofia
Jahrbuch der Land- und Forstwissenschaft, Fakultet der Universität in Sofia. Sofia. 1923–
Jahrb. Wiss. Bot.
Jahrbücher für wissenschaftliche Botanik. Berlin. 1858–1944.
Jahresb. Landes-Lehresem. Wiener-Neustadt
Jahresbericht des nieder-österrischischen Landes-Lehreseminars in Wiener-Neustadt. Wien.
Jahresb. Staats-ober-Realschule Steyr
Jahresbericht der K.K. Staats-ober-Realschule zu Steyr. Steyr.
Jap. J. Bot.
Japanese Journal of Botany. Tokyo. 1923–
Kansas Univ. Sci. Bull.
Kansas University Science Bulletin. Lawrence. 1902–
Kert. Föisk. Évkön.
Kertészeti és szölészeti föiskola évkönyve. Budapest. 1953–
Kew Bull.
Bulletin of miscellaneous information. Royal Gardens, Kew. London. 1887–1941. Continued as *Kew Bulletin*. London. 1946–
Kong. Danske Vid. Selsk.
Kongelige Danske Videnskabernes Selskabs Skrifter. København. 1745-1938
Kulturpflanze
Die Kulturpflanze. Berlin. 1953–
Kulturpflanze Beih.
Die Kulturpflanze Beihefte. Berlin. 1953–
Kung. Svenska Vet-Akad. Handl.
Kungliga Svenska Vetenskapsakademiens Handlingar. Uppsala and Stockholm. 1739–
Kurtziana
Kurtziana. Córdoba. 1961–
Leafl. West. Bot.
Leaflets of Western Botany. San Francisco. 1932–
Lilloa
Lilloa. Revista de botánica. Tucumán. 1937–
Linnaea
Linnaea. Halle. 1826–82.

Lunds Univ. Årsskr.
Lunds Universitets Årsskrift. Ser. 1. Vols 1–40. Lund. 1864–1902: ser. 2. Vols 1–. Lund. 1905–

Madroño
Madroño. Journal of the California Botanical Society. San Francisco, etc. 1916–

Mag. Zool. Bot.
Magazine of Zoology and Botany. Vols 1–2. London. 1837–38.

Magyar Bot. Lap.
Magyar Botanikai Lapok. Budapest. 1902–34.

Malpighia
Malpighia. Rassegna mensuale di Botanica. Messina and Genova. 1886–

Math. Term. Értesitö
Mathematikai és Természettudományi Értesitö. Budapest. 1882–1944.

Math. Term. Közl.
Mathematikai és Termeszettudományi Közlemények, Vonatkozálog a Hazai Viszonyokra. Budapest. 1861–1944.

Med. Landbouwhoogesch. Wageningen
Mededelingen van de Landbouwhoogeschool te Wageningen. Wageningen. 1908–

Meddel. Göt. Bot. Trad.
Meddelanden från Göteborgs botaniska Trädgard. Göteborg. 1924–66.

Meddel. om Grønl.
Meddelelser om Grønland, af Kommissionen for Ledelsen af de geologiske og geographiske Undersøgelsen in Grønland. København. 1879–

Mém. Acad. Roy. Sci. Lettr. Beaux-Arts Belg.
Nouveaux mémoires de l'Académie Royal des Sciences. Bruxelles. 1820–45. Continued as *Mémoires de l'Académie r. des sciences, des lettres et des beaux-arts de Belgique.* Bruxelles. 1847–

Mém. Acad. Sci. St. Pétersb. (Sci. Phys. Math.)
Mémoires de l'Académie Impériale des sciences de St. Pétersburg (classe des sciences, physiques et mathématiques). Sér. 5 Vols 1–11. St. Pétersburg. 1803–22: Sér 6 Vols 1–10. 1831–59: Sér. 7. Vols 1–42. 1859–97.

Mém. Acad. Toulouse
Mémoires de l'Académie des Sciences, Inscriptions et Belles-lettres de Toulouse. Toulouse. 1782–

Mem. Accad. Sci. Torino
Memorie della R. Accademia delle scienze di Torino. Torino. 1759–

Mem. Bot. Columbia Coll.
Memoirs from the Department of Botany of Columbia College. Vol. 1. Lancaster, Pennsylvania. 1895. Continued as *Memoirs from the Department of Botany of Columbia University.* Vol. 2. Lancaster, Pennsylvania. 1898.

Mem. Bot. Columbia Univ.
Memoirs from the Department of Botany of Columbia University. Vol. 2. Lancaster, Pennsylvania. 1898.

Mem. Cornell Univ. Agr. Exp. Stat.
Memoirs of the Cornell University Agricultural Experiment Station. Ithaca, New York. 1913–

Mém. Cour. Mém. Sav. Étr. Acad. Roy. Sci. Belg.
Mémoires couronnés et Mémoires des Savants étrangers. Académie royale des Sciences, des lettres et des beaux-arts de Belgique. Bruxelles. 1818–1904.

Mém. Herb. Boiss.
Mémoires de l'Herbier Boissier. Genève. 1900–

Mem. New York Bot. Gard.
Memoirs of the New York Botanical Garden. Vols 1–7. Bronx Park. 1900–27: vols 8–. Lancaster, Pennsylvania. 1952–

Mém. Soc. Acad. Maine Loire
Mémoires de la Société académique de Maine et Loire. Angers. 1857–

Mem. Soc. Brot.
Memórias da Sociedade broteriana. Coimbra. 1930–

Mem. Soc. Cienc. Nat. "La Salle"
Memorias de la Sociedad de Ciencias naturales "La Salle". Caracas. 1941–

Mém. Soc. Émul. Doubs
Mémoires de la Société d'Émulation du Doubs. Besançon. 1841–

Mem. Soc. Fauna Fl. Fenn.
Memoranda Societatis pro Fauna et Flora Fennica. Helsingforsiae. 1924–

Mém. Soc. Hist. Nat. d'Autun.
Mémoires de la Société d'Histoire naturelle d'Autun. 1888–

Mém. Soc. Imp. Sci. Nat. Cherbourg
Mémoires de la Société Impériale des Sciences naturelles. Cherbourg. 1852–

Mém. Soc. Philom.
Mémoires de la Société philomathique de Paris. Paris.

Mém. Soc. Phys. Nat. Genève
Mémoires de la Société de physique et d'histoire naturelle. Genève. 1821–

Mém. Soc. Roy. Sci. Lett. Arts Nancy
Mémoires de la Société r. des sciences, lettres et arts de Nancy. Nancy. 1831–48.

Mém. Soc. Sci. Hainaut
Mémoires et publications Société des sciences, des arts et des lettres du Hainaut. Mons. 1840–1920.

Mém. Soc. Sci. Nat. Phys. Maroc, Bot.
Mémoires de la Société des sciences naturelles et physiques du Maroc, sér. Botanique. Rabat and Paris. 1921–

Mem. Torrey Bot. Club
Memoirs of the Torrey Botanical Club. New York. 1889–

Mem. Wern. Nat. Hist. Soc.
Memoirs of the Wernerian Natural History Society. Vols 1–8. Edinburgh. 1811–38 [39].

Milwaukee Public Mus. Publ. Bot.
Milwaukee Public Museum Publications in Botany. Milwaukee. 1955–

Minn. Bot. Stud.
Minnesota Botanical Studies. St. Paul. 1894–1916.

Mitt. Basle. Bot. Ges.
Mitteilungen der Basler botanischen Gesellschaft. Basel. 1953–54. Continued as *Bauhinia*. Basel. 1955–

Mitt. Bot. Staats. München
Mitteilungen aus der botanischen Staatssammlung. München. 1950–

Mitt. Deutsch. Dendr. Ges.
Mitteilungen der deutschen dendrologischen Gesellschaft. Berlin, etc. 1893–

Mitt. Fl.-Soz. Arbeitsgem.
Mitteilungen der Floristisch-soziologischen Arbeitsgemeinschaft. Vols 1–2. Stolzenau. 1937: new ser. Vol. 1–. 1949–

Mitt. Thür. Bot. Ges.
Mitteilungen des thüringischen botanischen Gesellschaft. Weimar. 1949–

Mitt. Thür. Bot. Ver.
Mitteilungen des thüringischen botanischen Vereins. Weimar. 1882–1943. Continued as *Mitteilungen der thüringischen botanischen Gesellschaft*. Weimar. 1949–

Monatsber. Koenigl. Akad. (Berlin)
Monatsberichte der Koeniglich-preussischen Akademie der Wissenschaften. Berlin. 1856–81.

Monogr. Bot.
Monographiae botanicae. Warszawa. 1953–

Nat.
Naturalist. Hull, London, etc. 1864–

Nat. Belge
Le Naturaliste belge. Bruxelles. 1920–

Nat. Camb.
Nature in Cambridgeshire. Cambridge. 1958–

Nat. Canad.
Le Naturaliste Canadien. Québec. 1868–

Nat. Monsp.
Naturalia Monspeliensia, Série botanique. Montpellier. 1955–

Nederl. Kruidk. Arch.
Nederlandsch Kruidkundig Archief. Leiden. 1846–1951.

Neue Denkschr. Allg. Schweiz. Ges. Naturw.
Neue Denkschriften der Allgemeinen schweizerischen Gesellschaft für die gesamten Naturwissenschaften. Zürich. 1837–1906. Continued as *Neue Denkschriften der schweizerischen naturforschenden Gesellschaft.* 1907–18: continued as *Denkschriften der schweizerischen naturforschenden Gesellschaft.* 1920–

New Biology
New Biology. London and New York. 1945–60.

New Flora and Silva
New Flora and Silva. London. 1928–40.

New Phyt.
New Phytologist. Cambridge. 1902–

Not. Syst. (Leningrad)
Notulae systematicae ex Herbario Horti botanici Petropolitani/Notulae systematicae ex Herbario Horti Botanici Reipublicae Rossicae. Leningrad. 1919–26. Continued as *Notulae systematicae ex Herbario Instituti botanici Akademiae Scientiarum URSS.* Leningrad. Continued as *Notulae systematicae ex Herbario Instituti botanici Nomine V. L. Komarovii Academiae Scientiarum URSS.*– vol. 22. Mosqua and Leningrad. –1963.

Notes Roy. Bot. Gard. Edinb.
Notes from the Royal Botanic Garden, Edinburgh. Edinburgh. 1900–

Notizbl. Bot. Gart. Berlin
Notizblatt des Königlichen botanischen Gartens und Museums zu Berlin. Berlin-Dahlem. 1895–

Notul. Syst.
Notulae systematicae. Lisboa.

Nouv. Arch. Mus. (Paris)
Nouvelles archives du Muséum d'histoire naturelle. Paris. 1865–1914.

Nouv. Mém. Acad. Roy. Sci. Bel. Lettr. Brux.
Nouveaux mémoires de la Académie Royal des Sciences. Bruxelles. 1820–45. Continued as *Mémoires de l'Academie r. des sciences, des lettres et des beaux-arts de Belgique.* Bruxelles. 1847–

Nouv. Mém. Soc. Nat. Moscou
Nouveaux mémoires de la Société impériale des Naturalistes de Moscou. Moscou. 1829–

Nov. Bot. Hort. Bot. Univ. Prag.
Novitates botanicae ex Instituto et Horto Botanico Universitates Carolinae Pragensis. Praha. 1958–

Növ. Közlem.
Növénytani Közlemények. Budapest. 1902–8.

Nov. Syst. Plant. Vasc.
Novitates Systematicae Plantarum Vascularium. Academia Scientiarum U.R.S.S. Institutum Botanicum Nomini V. L. Komorovii. Mosqua and Leningrad. 1964–

Nova Acta Acad. Leop.-Carol.
Nova Acta (Leopoldina) Academiae Caesareae. Leopoldino-Carolinae germanicae Naturae curiosorum. Norimbergae, etc. 1757–1928: Nov. Ser. 1934–

Nova Acta Regiae Soc. Sci. Upsal.
Nova Acta Regiae Societatis scientiarum upsaliensis. Uppsala. 1773–

Nova Hedw.
Nova Hedwigia. Zeitschrift für Kryptogamenkunde. Weinheim. 1959–

Nucleus
Nucleus. International Journal of Cytology and Allied Subjects. Calcutta. 1958–

Nuov. Giorn. Bot.
Nuovo Giornale botanico italiano. Vols 1–68. Firenze. 1869–1966.

Nytt Mag. Nat.
Nyt magazin för naturvinskaberne. Vols 1–74. Christiana and Oslo. 1862–1925. Continued as *Nytt magasin för naturvidenskaberne.* Vols 75–88. Oslo. 1936–51. Continued as *Nytt magasin för botanik.* Vol. 1–. Oslo. 1952–

N.Z. J. Bot.
New Zealand Journal of Botany. Wellington. 1963–

Opera Bot.
Opera botanica. A Societate botanica Lundensi in Supplementum Seriei "Botaniska Notiser" edita. Lund. 1953–

Opera Bot. Čech.
Opera botanica čechica. Praha. 1940–46.

Österr. Bot. Zeitschr.
Österreichische botanische Zeitschrift. Wien. 1858–

Palestine J. Bot.
Palestine Journal of Botany. Jerusalem. 1938–54.

Papers Michigan Acad.
Papers from the Michigan Academy of Science, Art and Letters. New York. 1921–

Philipp. J. Sci. (Bot.)
Philippine Journal of Science. Botany. Manila. 1906–

Phyt.
Phytologist: a popular botanical miscellany. London. 1841–54: new series: *a botanical journal.* London. 1855–63.

Phytologia
Phytologia. New York. 1933–

Phyton
Phyton. Anngles Rei botanicae. Horn. 1948–

Planta
Planta. Archiv. für Wissenschaftliche Botanik. Berlin-Dahlem. 1925–

Planta Medica
Planta Medica. Stuttgart. 1953–

Plant Life
Plant Life. Stanford, California. 1945–

Pollichia
Pollichia. Ein naturwissenschaftlicher Verein der Rheinpfalz. 1843–1919. Continued as *Pfälzischer Verein für Naturkunde, Pollichia*. Durkheim a.d.H. 1920–

Prace Kom. Biol. (Poznań)
Prace Komisji Matematyczno Przyrodniczej, ser. B. Nauki Biologiczne. Vols 1–14. Poznań. 1921–48. Continued as *Prace Komisji Biologicznej*. Vol. 15–. Poznań. 1949–

Preslia
Preslia. Véstnik Česke (Československé) botanické Společnosti. Praha. 1914–

Proc. Amer. Acad. Arts Sci.
Proceedings of the American Academy of Arts and Sciences. Boston. 1846–1958. Continued as *Daedalus. Proceedings of the American Academy of Arts and Sciences*. 1958–

Proc. Bot. Soc. Brit. Isles
Proceedings of the Botanical Society of the British Isles. Arbroath. 1954–

Proc. Calif. Acad. Sci.
Proceedings of the California Academy of Sciences. San Francisco. 1854–

Proc. Cambridge Phil. Soc.
Proceedings of the Cambridge Philosophical Society. Cambridge. 1843–

Proc. Iowa Acad. Sci.
Proceedings of the Iowa Academy of Science. Des Moines. 1887–

Proc. Linn. Soc.
Proceedings of the Linnean Society of London. London. 1838–

Proc. Linn. Soc. New South Wales
Proceedings of the Linnean Society of New South Wales. Sydney. 1875–

Proc. Roy. Dublin Soc.
Scientific Proceedings of the Royal Dublin Society. Dublin. 1877–

Proc. Roy. Irish Acad.
Proceedings of the Royal Irish Academy. Dublin. 1836–

Programm K. K. ober-Rezlschule Innsbruck
Programm der K.K. ober-Rezlschule Innsbruck. Innsbruck. 1888–89.

Publ. Arnold Arb.
Publications of the Arnold Arboretum. Cambridge, Mass. 1891–

Publ. Fac. Sci. Univ. Masaryk (Brno)
Publications de la Faculté des sciences de l'Université Masaryk. Brno. 1921–58.

Publ. Field Mus. Nat. Hist. Bot. Ser.
Publications. Field Museum of Natural History. Chicago. 1895–1947. Continued as *Fieldiana: Botany*. 1948–

Quart. Bull. Alpine Gard. Soc.
Quarterly Bulletin of the Alpine Garden Society. London. 1933–

Rec. Trav. Bot. Néerl.
Receueil des Travaux Botaniques Néerlandais. Société Botanique Néerlandaise. Nimègue 1904–50.

Regn. Veg.
Regnum Vegetabile. A series of publications for the use of plant taxonomists and plant geographers. Utrecht. 1952–

Rep. Bot. Soc. & E.C.
Report of the Botanical Society and Exchange Club of the British Isles. London, Manchester, Oxford and Arbroath. 1867–1948.

Rep. Missouri Bot. Gard.
Report of the Missouri Botanical Garden. St. Louis. 1890–1912. Continued as *Annals of the Missouri Botanical Garden.* St. Louis. 1914–

Rep. Welsh Plant Breed. Stat.
Report. Welsh Plant Breeding Station. Aberystwyth. 1931–

Res. Stud. State Coll. Washington
Research Studies, State College of Washington, later Research Studies, Washington State University. Pullman. 1929–

Rev. Bot.
Revue de Botanique. Bulletin mensuel de la Société française de Botanique. Toulouse. 1882–95.

Rev. Bot. Syst. Géogr. Bot.
Revue de Botanique Systématique et de Géographie Botanique. Vols 1–2. Tours. 1903–5.

Rev. Roum. Biol.
Revue Roumaine de Biologie: Série botanique. Bucharest. 1956–

Rev. Sci. Nat.
Revue des Sciences naturelles. Montpellier and Paris. 1872–85.

Rhodora
Rhodora: Journal of the New England Botanical Club. Boston. 1899–

Rozpr. České Akad. Věd Umeni
Rozpravy České Akademie věd a Umeni. Třida 2 (*Matematicko-přirodovědecká*). Praha. 1921–52.

Sborn. Bălg. Akad. Nauk.
Sbornik na Bălgarskaga Akademija na Naukitě. Sofija. 1911–

Sborn. národ. Mus. Praha
Sbornik národniho musea v Praze. Praha. 1938–

Scripta Hier.
Scripta Hierosolymitana. Jerusalem. 1954–

Sitz.-Ber. Akad. Wiss. Wien
Sitzungsberichte der Kaiserlichen Akademie der Wissenschaften in Wien. Mathematisch-naturwissenschaftliche Klasse. Wien. 1848–1947. Continued as *Sitzungsberichte der Österreischen Akademie der Wissenschaften Mathematisch-naturwissenschaftliche Klasse.* Wien. 1947–

Sitz. Österr. Akad. Wiss. Math.-Nat. Kl.
Sitzungsberichte der Österreichischen Akademie der Wissenschaften Mathematisch-naturwissenschaftliche Klasse. Wien. 1947–

Southwest. Nat. (Texas)
Southwestern Naturalist. Dallas. 1956–

Sov. Bot.
Sovietskaya Botanica. Leningrad. 1933–

Spisy Přírod. Univ. Purkyně Brně
Spisy Přírodovědecké Fakulty University J. E. Purkyně v Brně. Brno. 1959–

Svensk Bot. Tidskr.
Svensk botanisk Tidskrift. Stockholm. 1907–

Symb. Bot. Upsal.
Symbolae botanicae Upsaliensis. Arbeten från botaniska Institutionen i Uppsala. Uppsala. 1932–

Taxon
Taxon. International Association for Plant Taxonomy. Utrecht. 1951–

Theses Fac. Sci. Toulouse
Theses. Faculté des Sciences, Université de Toulouse. Toulouse.

Trans. Acad. Sci. St. Louis
Transactions of the Academy of Science, St. Louis. St. Louis. 1856–

Trans. Linn. Soc.
Transactions of the Linnean Society. Ser. 1 Vols 1–6. London. 1791–1802. Continued as *Transactions of the Linnean Society of London,* ser. 1 Vols 7–30. 1804–75. Continued as *Transactions of the Linnean Society of London: Botany,* ser. 2 Vol. 1–. 1875–. From 1939 *Botany* and *Zoology* are combined.

Trans. Proc. Bot. Soc. Edinb.
Transactions and Proceedings of the Botanical Society of Edinburgh. Edinburgh. 1841–

Trans. Roy. Soc. Canada
Transactions of the Royal Society of Canada. Ottawa. 1882–

Trans. Roy. Soc. Edinb.
Transactions of the Royal Society of Edinburgh. Edinburgh. 1783–

Trans. Roy. Soc. N.Z.
Transactions of the Royal Society of New Zealand. Dunedin. 1952–

Trav. Inst. Sci. Chérif.
Travaux de l'Institut chérifien. Tanger. Continued as *Travaux de l'Institut scientifique chérifien*. Tanger. 1951–

Trav. Inst. Sci. Nat. Peterhoff
Travaux de l'Institut des Sciences naturelles de Peterhoff. Peterhof. 1925–32.

Trav. Lab. Biol. Vég. Grenoble-Lautaret
Travaux du Laboratorie de biologie et végetale. Grenoble-Lautaret.

Trav. Mus. Bot. Acad. Pétersb.
Travaux du Musée botanique de l'Académie Impériale des Sciences de St. Pétersburg (de l'Académie Impériale des Sciences de Russie). St. Pétersburg. 1902–32.

Trommsd. N.J. Pharm.
Trommsdorff Journal für Pharmacie, title later changed to *Neues Journal für Pharmacie*. Erfurt. 1794–1834.

Trudý Inst. Biol. Riga
Trudý Instituta biologii Akademiya nauk Latviĭskoĭ SSSR. Riga. 1953–

Univ. Calif. Publ. Bot.
University of California Publications in Botany. Berkeley. 1902–

Univ. Colorado Stud.
University of Colorado Studies. Boulder. 1902–

Univ. Washington Publ. Biol.
University of Washington Publications in Biology. Seattle. 1932–

U.S. Dept. Agric. Agrost. Bull.
United States Department of Agriculture, Division of Agrostology Bulletin. Washington. 1895–1901.

U.S. Dept. Agric. Hand.
United States Department of Agriculture. Handbook. Washington. 1944–

U.S. Dept. Agric. Techn. Bull.
United States Department of Agriculture Technical Bulletin. Washington. 1927–

Verh. Bot. Ver. Brandenburg
Verhandlungen des botanischen Vereins der Provinz Brandenburg. Berlin. 1859–

Verh. Kaiser Franz Josef Landes-Gymnasium Baden bei Wien
Verhandlungun der Kaiser Franz Josef Landes-Gymnasium. Baden bei Wien.

Verh. Kon. Akad. Wetenschapp.
Verhandelingen der K. akademie van wetenschappen. Amsterdam. 1892–1937. Continued as *Verhandelingen der K. nederlandsche akademie van wetenschappen*. Amsterdam. 1937–

Verh. Naturf. Ges. Basel
Verhandlungen der naturforschenden Gesellschaft in Basel. Basel. 1854–

Verh. Zool.-Bot. Ges. Wien

Verhandlungen der K.K. zoologisch-botanischen Verein in Wien. Wien. 1852–. From 1858 *Verein* is replaced by *Gesellschaft.*

Veröff. Geobot. Inst. Rübel

Veröffentlichungen des geobotanischen Institutes Rübel. Zürich. 1924–

Viert. Naturf. Ges. Zürich

Vierteljahrsschrift der naturforschenden Gesellschaft in Zürich. Zürich. 1856–

Watsonia

Watsonia. Journal of the Botanical Society of the British Isles. Arbroath, Salisbury and Ashford. 1949–

Webbia

Webbia. Raccolta di scritti botanici. Firenze. 1905–

Weeds

Weeds. Association of Regional Control Conferences. New York, etc. 1951–

Wentia

Wentia. Supplement to *Acta Botanica Neerlandica.* Amsterdam. 1959–

Werdenda

Werdenda. Beiträge zu Pflanzenkunde. Bingen, Washington. 1923–31.

Willdenowia

Willdenowia. Mitteilungen aus dem Botanischen Garten und Museum Berlin-Dahlem. Berlin. 1953–

Wiss. Zeitschr. Univ. Halle Math.-Nat.

Wissenschaftliche Zeitschrift der Martin-Luther-Universität, Halle-Wittenberg: Mathematisch-naturwissenschaftliche Reihe. Halle-Wittenberg. 1953–

Wyoming Univ. Publ. Bot.

University of Wyoming Publications in Science, later *University of Wyoming Publications: Botany.* Laramie. 1922–

Zeitschr. Indukt. Abst.-Vererb.

Zeitschrift für induktive Abstammungs-u.-Vererbungslehre. Berlin. 1908–43, 1948–57.

Züchter

Der Züchter. Internationale Zeitschrift für theoretische und angewandte Genetik. Berlin. 1929–

Index to Botanical Monographs

PTERIDOPHYTA

Bower, F. O. *The Ferns (Filicales)*. Vol. 1. Pp. 359. Cambridge. 1923: vol. 2. Pp. 344. 1926: vol. 3. Pp. 306. 1928.

Hooker, Sir W. J. *Species Filicum*. Vol. 1. Pp. 245. London. 1846: vol. 2. Pp. 250. 1858: vol. 3. Pp. 291. 1860: vol. 4. Pp. 292. 1862: vol. 5. Pp. 314. 1864.

Hooker, Sir W. J. and Bauer, F. *Genera Filicum, or Illustrations of the Ferns*. London. 1842.

Hyde, H. A. and Wade, A. E. *Welsh Ferns*. Pp. x + 131 + 10 plates and 67 text figures. Cardiff. 1940. Edition 2. Pp. x + 131 + 10 plates and 67 text figures. 1948. Edition 3. Pp. x + 131 + 10 plates and 67 text figures. 1954. Edition 4. Pp. ix + 122 + 12 plates and 70 text figures. 1962.

Lowe, E. J. *Ferns British and Exotic*. 8 vols. London. 1856–60.

Lowe, E. J. *Our Native Ferns, or a History of the British species and their varieties*. Vol. 1. Pp. 348. London. 1879: vol. 2. Pp. 492. 1880.

Manton, I. *Problems of Cytology and Evolution in the Pteridophyta*. Pp. x + 316. Cambridge. 1950.

Milde, J. *Filices Europae et Atlantidis, Asiae minoris et Siberiae*. Pp. 311. 1867.

Newman, E. *A History of British Ferns*. Pp. xxxiv + 104. London. 1840. Edition 2. Pp. xxxii + 424. 1844. Edition 3. Pp. xvi + 364. 1854. Edition 4. Pp. 196. 1865.

Presl, C. B. *Tentamen Pteridographiae, seu Genera Filicacearum praesertim juxta Venarum Decursum et Distributionem exposita*. Pp. 290. Pragae. 1836. *Supplementum*. Pragae. 1845.

Swartz, O. *Synopsis Filicum, earum Genera et Species systematice complectens. Adjectis Lycopodineis et Descriptionibus novarum et rariorum Specierum*. Pp. 445. Kiliae. 1806.

Verdoorn, F. (Ed.). *Manual of Pteridology*. Pp. xx + 640. The Hague. 1938.

LYCOPODIACEAE

Nessel, H. *Die Bärlappgewächse (Lycopodiaceae)*. Pp. viii + 404. Jena. 1939.

Newman, E. A history of the British Lycopodia and allied genera. *Phyt.* **1**, 1–7, 18–20, 33–36, 49–51, 81–86. 1841.

Rothmaler, W. Pteridophyten-Studien, 1. *Fedde. Rep.* **54**, 55–82. 1944.

Spring, A. Monographie de la famille des Lycopodiacées. *Nouv. Mém. Acad. Roy. Sci. Bel. Lettr. Brux.* **15** (Mém. 6), 1–110. 1842: *Mém. Acad. Roy. Sci. Lettr. Beaux-Arts. Belg.* **24** (Mém. 1), 1–358. 1850.

***HUPERZIA* (*LYCOPODIUM*, p.p.)**

Herter, G. *Index Lycopodiorum.* Pp. 120. Montevideo. 1949.

Lloyd, F. and Underwood, L. M. Review of the species of Lycopodium of North America. *Bull. Torrey Bot. Club* **27**, 147–168. 1900.

Löve, A. and Löve, D. Cytotaxonomy and classification of Lycopods. *Nucleus* **1**, 1–10. 1958.

Nessel, H. Beiträge zur Kenntnis der Gattung Lycopodium. *Fedde. Rep.* **39**, 61–71. 1937.

Nessel, H. *Die Bärlappgewächse* (*Lycopodiaceae*). Pp. viii + 404. Jena. 1939.

Newman, E. A history of the British Lycopodia, and allied genera. *Phyt.* **1**, 1–7, 18–20, 33–36, 49–51, 81–86. 1841.

Rothmaler, W. Pteridophyten-Studien, 1. *Fedde. Rep.* **54**, 55–82. 1944.

Spring, A. Monographie de la famille des Lycopodiacées. *Nouv. Mém. Acad. Roy. Sci. Bel. Lettr. Brux.* **15** (Mém. 6), 1–110. 1842: *Mém. Acad. Roy. Sci. Lettr. Beaux-Arts Belg.* **24** (Mém. 1), 1–358. 1850.

Wilson, L. R. The spores of the genus Lycopodium in the United States and Canada. *Rhodora* **36**, 13–19. 1934.

***LYCOPODIELLA* (*LYCOPODIUM*, p.p.: *LEPIDOTIS*)**

Herter, G. *Index Lycopodiorum.* Pp. 120. Montevideo. 1949.

Lloyd, F. and Underwood, L. M. Review of the species of Lycopodium of North America. *Bull. Torrey Bot. Club* **27**, 147–168. 1900.

Löve, A. and Löve, D. Cytotaxonomy and classification of Lycopods. *Nucleus* **1**, 1–10. 1958

Nessel, H. Beiträge zur Kenntnis der Gattung Lycopodium. *Fedde. Rep.* **39**, 61–71. 1937.

Nessel, H. *Die Bärlappgewächse* (*Lycopodiaceae*). Pp. viii + 404. Jena. 1939.

Newman, E. A history of the British Lycopodia, and allied genera. *Phyt.* **1**, 1–7, 18–20, 33–36, 49–51, 81–86. 1841.

Rothmaler, W. Pteridophyten-Studien, 1. *Fedde. Rep.* **54**, 55–82. 1944.

Spring, A. Monographie de la famille des Lycopodiacées. *Nouv. Mém. Acad. Roy. Sci. Bel. Lettr. Brux.* **15** (Mém. 6), 1–110. 1842: *Mém. Acad. Roy. Sci. Lettr. Beaux-Arts Belg.* **24** (Mém. 1), 1–358. 1850.

Wilson, L. R. The spores of the genus Lycopodium in the United States and Canada. *Rhodora* **36**, 13–19. 1934.

LYCOPODIUM

Herter, G. *Index Lycopodiorum.* Pp. 120. Montevideo. 1949.

Lloyd, F. and Underwood, L. M. Review of the species of Lycopodium of North America. *Bull. Torrey Bot. Club* **27**, 147–168. 1900.

Löve, A. and Löve, D. Cytotaxonomy and classification of Lycopods. *Nucleus* **1**, 1–10. 1958.

Nessel, H. Beiträge zur Kenntnis der Gattung Lycopodium. *Fedde. Rep.* **39**, 61–71. 1937.

Nessel, H. *Die Bärlappgewächse* (*Lycopodiaceae*). Pp. viii + 404. Jena. 1939.

Newman, E. A history of the British Lycopodia, and allied genera. *Phyt.* **1**, 1–7, 18–20, 33–36, 49–51, 81–86. 1841.

Rothmaler, W. Pteridophyten-Studien, 1. *Fedde. Rep.* **54**, 55–82. 1944.

Spring, A. Monographie de la famille des Lycopodiacées. *Nouv. Mém. Acad. Roy. Sci. Bel. Lettr. Brux.* **15** (Mém. 6), 1–110. 1842: *Mém. Acad. Roy. Sci. Lettr. Beaux-Arts Belg.* **24** (Mém. 1), 1–358. 1850.

Wilson, L. R. The spores of the genus Lycopodium in the United States and Canada. *Rhodora* **34**, 13–19. 1934.

***DIPHASIUM* (*LYCOPODIUM*, p.p.)**

Herter, G. *Index Lycopodiorum.* Pp. 120. Montevideo. 1949.

Lloyd, F. and Underwood, L. M. Review of the species of Lycopodium of North America. *Bull. Torrey Bot. Club* **27**, 147–168. 1900.

Löve, A. and Löve, D. Cytotaxonomy and classification of Lycopods. *Nucleus* **1**, 1–10. 1958.

Nessel, H. Beiträge zur Kenntnis der Gattung Lycopodium. *Fedde. Rep.* **39**, 61–71. 1937.

Nessel, H. *Die Bärlappgewächse* (*Lycopodiaceae*). Pp. viii + 404. Jena. 1939.

Newman, E. A history of the British Lycopodia, and allied genera. *Phyt.* **1**, 1–7, 18–20, 33–36, 49–51, 81–86. 1841.

Rothmaler, W. Pteridophyten-Studien, 1. *Fedde. Rep.* **54**, 55–82. 1944.

Rothmaler, W. Über einige Diphasium-Arten (Lycopodiaceae). *Fedde. Rep.* **66**, 234–236. 1962.

Spring, A. Monographie de la famille des Lycopodiacées. *Nouv. Mém. Acad. Roy. Sci. Bel. Lettr. Brux.* **15** (Mém. 6), 1–110. 1842: *Mém. Acad. Roy. Sci. Lettr. Beaux-Arts Belg.* **24** (Mém. 1), 1–358. 1850.

Wilce, J. H. Section Complanata of the genus Lycopodium. *Beih. Nova Hedw.* **19**. Pp. 233 + 15 tables + 40 plates. 1965.

Wilson, L. R. The spores of the genus Lycopodium in the United States and Canada. *Rhodora* **36**, 13–19. 1934.

SELAGINELLACEAE

SELAGINELLA

Alston, A. H. G. The Heterophyllous Selaginellae of Continental North America. *Bull. Brit. Mus.* (*Bot.*) **1**, 221–274 + 2 plates. 1955.

Baker, J. G. A synopsis of the genus Selaginella. *J. Bot.* **21**, 1–5, 42–46, 80–84, 97–100, 141–145, 210–213, 240–244, 332–336. 1883: *loc. cit.* 22,

23–26, 86–90, 110–113, 243–247, 275–278, 295–300, 373–377. 1884: *loc. cit.* 23, 19–25, 45–48, 116–122, 154–157, 176–180, 248–252, 292–302. 1885.

CHOISY, J. D. Mémoire sur la famille des Sélaginées. *Mém. Soc. Phys. Nat. Genève* **2**, 71–114 + 5 tables. 1824.

CLAUSEN, R. T. Selaginella subgenus Euselaginella in the south-eastern United States. *Amer. Fern J.* **36**, 65–82. 1946.

NEWMAN, E. A history of the British Lycopodia, and allied genera. *Phyt.* **1**, 1–7, 18–20, 33–36, 49–51, 81–86. 1841.

REED, C. F. Index Selaginellarum. *Mem. Soc. Brot.* **18**, 1–287. 1966.

REEVE, R. M. The spores of the genus Selaginella in north, central and northeastern United States. *Rhodora* **37**, 342–345. 1935.

ROTHMALER, W. Pteridophyten-Studien, 1. *Fedde. Rep.* **54**, 55–82. 1944.

SPRING, A. Monographie de la famille des Lycopodiacées. *Nouv. Mém. Acad. Roy. Sci. Bel. Lettr. Brux.* **15** (Mém. 6), 1–110. 1842: *Mém. Acad. Roy. Sci. Lettr. Beaux-Arts Belg.* **24** (Mém. 1), 1–358. 1850.

ISOETACEAE

ISOETES

BAKER, J. G. A synopsis of the species of Isoetes. *J. Bot.* **18**, 65–70, 105–110. 1880.

ENGELMANN, G. The genus Isoetes in North America. *Trans. Acad. Sci. St. Louis* **4**, 358–390. 1882.

FUCHS, H. P. Nomenklatur, Taxonomie und Systematik der Gattung Isoëtes Linnaeus in geschichtlicher Betrachtung. *Beih. Nova Hedw.* **3**. Pp. 103 + 13 plates. 1962.

GRENDA, A. Ueber die Systematische Stellung der Isoëtaceen. *Bot. Arch.* **16**, 268–296. 1926.

HY, F. Les Isoetes amphibies de la France centrale. *J. Bot. (Paris)* **8**, 92–98. 1894.

LÖVE, A. Cytotaxonomy of the Isoetes echinospora complex. *Amer. Fern J.* **52**, 113–123. 1962.

*MOTELAY, L. and VENDRYÉS, H. Monographie des Isoëtaceae. *Act. Soc. Linn. Bordeaux* **36**. Pp. 96. 1882.

PFEIFFER, N. E. Monograph of the Isoetaceae. *Ann. Missouri Bot. Gard.* **9**, 79–232. 1922.

REED, C. F. Index Isoëtales. *Bol. Soc. Brot. ser.* 2 **27**, 5–72. 1953.

ROTHMALER, W. Pteridophyten-Studien, 1. *Fedde. Rep.* **54**, 55–82. 1944.

EQUISETACEAE

EQUISETUM

DUVAL-JOUVE, J. *Histoire naturelle des Equisetum de France.* Pp. viii + 284. Paris. 1864.

HAUKE, R. L. A resumé of the taxonomic reorganization of Equisetum subgenus Hippochaete. *Amer. Fern J.* **51**, 131–137. 1961: *loc. cit.* **52**, 29–35, 57–63, 123–130. 1962.

HAUKE, R. L. A taxonomic monograph of the genus Equisetum subgenus Hippochaete. *Beih. Nova Hedw.* **8**. Pp. 1–123 + 12 tables + 3 graphs + 22 plates. 1963.

KUMMERLE, J. B. Equisetum-Bastarde als verkannte Artformen. *Magyar Bot. Lap.* **30**, 146–160. 1931.

MILDE, J. *Monographia Equisitorum.* Pp. 605 + 35 plates. Dresden. 1865.

NEWMAN, E. A history of the British Equiseta. *Phyt.* **1**, 274–278, 305–308, 337–340. 1842: *loc. cit.* **1**, 529–535, 593–594, 627–630, 689–700, 721–731. 1843.

ROSENDAHL, H. V. De svenska Equisetum-arterna och deras former. *Arkiv Bot.* **15** (3). Pp. 52. 1918.

ROTHMALER, W. Pteridophyten-Studien, 1. *Fedde. Rep.* **54**, 55–82. 1944.

SCHAFFNER, J. H. Studies of Equiseta in European herbaria. *Amer. Fern J.* **21**, 90–102. 1931.

VAUCHER, J. P. É. Monographie des Prêles. Histoire générale et physiologique du genre. *Mém. Soc. Phys. Nat. Genève* **1**, 329–391. 1822.

PTEROPSIDA

CHRIST, H. *Die Farnkräuter der Erde.* Pp. xii + 388. Jena. 1897.

CHRISTENSEN, C. *Index Filicum.* Pp. lix + 744. Hafniae. 1906. Supplementum, 1906–1912. Pp. 131. 1913. Supplément Préliminaire. Pp. 60. 1917. Supplementum Tertium, pro annis 1917–1933. Pp. 219. 1934. Supplementum Quartum, pro annis 1934–1960, edited by R. E. G. Pichi-Sermolli. *Regn. Veg.* **37**. Pp. xiv + 370. 1965.

FEE, A. L. A. *Mémoires sur la Famille des Fougères.* 11 vols. Strasbourg. 1845–66.

HOOKER, Sir WILLIAM J. and BAKER, J. G. *Synopsis Filicum: or, A Synopsis of all known ferns.* Pp. 482. London. 1865–68. Edition 2. Pp. 559. London. 1874.

SCHOTT, H. W. *Genera Filicum.* Vindobonae. 1834.

OSMUNDACEAE

OSMUNDA

MILDE, J. *Monographia Generis Osmundae.* Pp. 140. Vindobonae. 1868.

HYMENOPHYLLACEAE

TRICHOMANES

COPELAND, E. B. Trichomanes. *Philipp. J. Sci. (Bot.)* **51**, 119–280. 1933.

HYMENOPHYLLUM

COPELAND, E. B. Genera Hymenophyllacearum. *Philipp. J. Sci.* (*Bot.*) **67**, 1–110. 1938.

EVANS, G. B. The identification of British Hymenophyllum species. *Brit. Fern Gaz.* **9**, 256–262. 1966.

METTENIUS, G. Über die Hymenophyllaceae. *Abh. Sächs. K. Ges. Wiss.* **7**, 403–504 + Taf. 5. 1864.

PRESL, C. B. Hymenophyllaceae. *Abh. Böhm. Ges. Wiss.* (*Math.-Nat.*) **5** (3). Pp. 93. 1843.

VAN DEN BOSCH, R. B. Synopsis Hymenophyllacearum *Nederl. Kruidk. Arch.* **4**, 341–419. 1859.

DENNSTAEDTIACEAE

PTERIDIUM

AGARDH, J. *Recensio Specierum Generis Pteridis.* Pp. iv + 86. Lundae. 1839.

BRAID, K. W. Bracken: a review of the literature. *Comm. Bur. Past. Field Crops Mim. Publ.* **3**. 1959.

TRYON, R. M. Jr. A revision of the genus Pteridium. *Rhodora* **43**, 1–31, 37–67. 1941.

PTERIS

SHIEH, W.-C. A synopsis of the fern genus Pteris in Japan. *Bot. Mag.* (*Tokyo*) **79**, 283–292. 1966.

ADIANTACEAE

ADIANTUM

*KEYSERLING, A. Gen. Adiantum. *Mém. Acad. Sci. St. Pétersb.* (*Sci. Phys. Math.*) *sér.* 7 **7**. 1875.

KUHN, M. Übersicht über die Arten der Gattung Adiantum. *Jahrb. Berlin Bot. Gart.* **1**, 337–351. 1881.

ASPLENIACEAE

ASPLENIUM

ALSTON, A. H. G. Notes on the supposed hybrids in the genus Asplenium found in Britain. *Proc. Linn. Soc.* **152**, 132–144. 1940.

CHRIST, H. Die Asplenien des Heufler'schen Herbars. *Allgem. Bot. Zeitschr.* **9**, 1–4, 28–31. 1904.

HEUFLER, L. R. von. Asplenii Species Europeae. *Verh. Zool.-Bot. Ges. Wien* **6**, 235–254. 1856.

LOVIS, J. D. The taxonomy of Asplenium trichomanes in Europe. *Brit. Fern Gaz.* **9**, 147–160. 1964.

MANTON, I. and REICHSTEIN, T. Diploides Asplenium obovatum Viv. *Bauhinia* **2**, 79–91. 1962.

MEYER, D. E. Untersuchungen über Bastardieuring in der Gattung Asplenium. *Bibl. Bot.* **123**, 1–34. 1952.

MEYER, D. E. Zur Zytologie der Asplenien Mittel-europas (I-XV). *Ber. Deutsch. Bot. Ges.* **70**, 57–66. 1957: (XVI–XX). *loc. cit.* **71**, 11–20. 1958: (XXI bis–XXIII). *loc. cit.* **72**, 37–48. 1959: (XXIV–XXVIII). *loc. cit.* **73**, 368–394. 1960: (XXIX. Abschluss). *loc. cit.* **74**, 449–461. 1961.

MEYER, D. E. Hybrids in the genus Asplenium found in north-western and central Europe. *Amer. Fern J.* **50**, 138–145. 1960.

MEYER, D. E. Über Typus-Exemplare von Asplenium-Bastarden Mittel-europas. *Willdenowia* **2**, 519–531. 1960.

MEYER, D. E. Über neue und seltene Asplenien Europas. *Ber. Deutsch. Bot. Ges.* **75**, 24–33. 1962: *loc. cit.* **76**, 13–22. 1963: *loc. cit.* **77**, 3–13. 1964.

ATHYRIACEAE

ATHYRIUM

MILDE, J. Das Genus Athyrium. *Bot. Zeit.* **24**, 373–376. 1866.

CYSTOPTERIS

BLASDELL, F. A monographic study of the fern genus Cystopteris. *Mem. Torrey Bot. Club* **21** (4), 1–102. 1963.

MUSSACK, A. Untersuchungen über Cystopterisfragilis. *Beih. Bot. Centr.* **51** (1), 204–254. 1933.

WOODSIA

BROWN, D. F. M. A monographic study of the fern genus Woodsia. *Beih. Nova Hedw.* **16**. Pp. x + 154 + 40 plates. 1964.

BROWN, R. On Woodsia, a new genus of ferns. *Trans. Linn. Soc.* **11**, 170–174. 1813.

PICHI-SERMOLLI, R. E. G. Il genere Woodsia R.Br. in Italia. *Webbia* **12**, 179–216. 1956.

POELT, J. Zur Kenntnis der Gattung Woodsia in Europa. *Mitt. Bot. Staats. München* **1**, 167–174. 1952.

ASPIDIACEAE

DRYOPTERIS

CHRISTENSEN, C. *On a natural classification of the species of Dryopteris*, in Rosenvinge, L. K., *Biologiske arbejder tilegnede Eug. Warming paa hans 70 aars födseldag den 3 November* 11, 73–85. København. 1911.

CRANE, F. W. Comparative study of diploid and tetraploid spores of Dryopteris dilatata from Britain and Europe. *Watsonia* **3**, 168–169. 1955.

CRANE, F. W. Spore studies in Dryopteris. *Amer. Fern J.* **43**, 159–169. 1953: *loc. cit.* **45**, 14–16. 1955: *loc. cit.* **46**, 127–130. 1956.

KARPOWICZ, W. Studium porównawcze nad krajowymi przedstawicielami gatunków rodzaju Dryopteris i Thelypteris. *Acta Soc. Bot. Pol.* **29**, 175–189. 1960.

WALKER, S. Cytogenetic studies in the Dryopteris spinulosa complex. *Watsonia* **3**, 193–209. 1955: *Amer. J. Bot.* **48**, 607–614. 1961.

WALKER, S. Cytotaxonomic studies of some American species of Dryopteris. *Amer. Fern J.* **49**, 104–112. 1959.

POLYSTICHUM

ELLIOT, E. A. The British Polystichums. *Brit. Fern Gaz.* **8**, 159 168. 1956.

MEYER, E. Zur Gattung Polystichum in Mitteleuropa. *Willdenowia* **2**, 336–342. 1960.

THELYPTERIDACEAE

THELYPTERIS

KARPOWICZ, W. Studium porównawcze nad krajowymi przedstawicielami gatunków rodzaju Dryopteris i Thelypteris. *Acta Soc. Bot. Pol.* **29**, 175–189. 1960.

POLYPODIACEAE

POLYPODIUM

BECKERS, B. Bijdrage tot de biosystematiek van Polypodium L. in België en het groothertogdom Luxemburg. *Bull. Jard. Bot. Bruxelles* **36**, 354–382. 1966.

MARTENS, P. Le paraphyses de Polypodium vulgare et la sous-espèce serratum. *Bull. Soc. Bot. Belg.* **82**, 225–262. 1950.

ROTHMALER, W. and SCHNEIDER, U. Die Gattung Polypodium in Europa. *Kulturpflanze Beih.* **3**, 234–248. 1962.

SHIVAS, M. G. Contributions to the cytology and taxonomy of the species of Polypodium in Europe and America, 1. Cytology. *J. Linn. Soc. Bot.* **58**, 13–24. 1961: 2. Taxonomy, *loc. cit.* **58**, 27–38. 1961.

AZOLLACEAE

AZOLLA

DI FULVIO, T. E. Sobre el episporio de las especies Americanas de Azolla con especial referencia a A. mexicana Presl. *Kurtziana* **1**, 299–302. 1961.

MARSH, A. S. Azolla in Britain and Europe. *Proc. Cambridge Phil. Soc.* **17**, 383–386. 1914. Reprinted in *J. Bot.* **52**, 209–213. 1914.

SVENSON, H. K. The New World species of Azolla. *Amer. Fern J.* **34**, 69–84. 1944.

OPHIOGLOSSACEAE

CLAUSEN, R. A monograph of the Ophioglossaceae. *Mem. Torrey Bot. Club* **19** (2), 1–177. 1938.

BOTRYCHIUM

CLAUSEN, R. A monograph of the Ophioglossaceae. *Mem. Torrey Bot. Club* **19** (2), 1–177. 1938.

MILDE, J. Monographia Botrychiorum. *Verh. Zool.-Bot. Ges. Wien* **1869**, 55–190. 1869.

OPHIOGLOSSUM

CLAUSEN, R. A monograph of the Ophioglossaceae. *Mem. Torrey Bot. Club* **19** (2), 1–177. 1938.

NISHIDA, M. Japanese species of Ophioglossum and their nomenclature. *J. Jap. Bot.* **34**, 33–47. 1959.

GYMNOSPERMAE

DALLIMORE, W. and JACKSON, A. B. *A Handbook of Coniferae*. London. 1923. Edition 2. 1931. Edition 3. 1948. Edition 4. Revised by S. G. Harrison. Pp. xix + 729. 1966.

DEBAZAC, E. F. *Manuel des Conifères*. Pp. 172. Nancy. 1965.

ENDLICHER, S. L. *Synopsis Coniferarum*. Pp. iv + 368. Sangalli. 1847.

FITZPATRICK, H. M. Conifers: keys to the genera and species. *Proc. Roy. Dublin Soc., new ser.* **19**, 189–260. 1929: edition 2, *loc. cit., ser. A* **2** (7), 67–129. 1965.

*FRANCO, J. do AMARAL. *De Coniferarum duarum Nominibus*. Lisboa. 1950.

OUDEN, P. den and BOOM, B. K. *Manual of Cultivated Conifers Hardy in the Cold- and Warm-Temperate Zone*. Pp. x + 526. The Hague. 1965.

PINACEAE

ABIES

FRANCO, J. do AMARAL. *Abetos*. Pp. vii + 260. Lisboa. 1950.

FULLING, E. H. Identification by leaf structure of the species of Abies cultivated in the United States. *Bull. Torrey Bot. Club* **61**, 497–524. 1934.

MATTFELD, J. Die in Europa und dem Mittelmeergebiet wildwachsenden Tannen. *Mitt. Deutsch. Dendr. Ges.* **1925**, 1–37. 1925.

MATZENKO, A. Conspectus generis Abies. *Not. Syst. (Leningrad)* **22**, 33–42. 1963.

M'NAB, W. R. A revision of the species of Abies. *Proc. Roy. Irish Acad. ser.* 2 *(Science)* **2**, 673–704. 1877.

LARIX

OSTENFELD, C. H. and LARSEN, C. S. The species of the genus Larix and their geographical distribution. *Biol. Medd.* **9** (2), 1–106. 1930.

PINUS

ENGELMANN, G. Revision of the genus Pinus and description of Pinus Elliottii. *Trans. Acad. Sci. St. Louis* **4**, 161–190 + 3 plates. 1880.

*GAUSSEN, H. *Le Gymnosperms*. Fasc. vi. ch. xi. *Pinus*. 1960.

HARLOW, W. M. The identification of the pines of the United States native and cultivated, by needle structure. *Bull. New York State Coll. Forestry* **4**, 1–21. 1931.

JÄHRIG, M. Beiträge zur Nadelanatomie und Taxonomie der Gattung Pinus L. *Willdenowia* **3**, 329–366. 1962.

*KLIKA, J., et al. *Jehličnaté.* Praha. 1953.

LAMBERT, A. B. *A description of the genus Pinus.* Pp. 189. 2 vols. London. 1803–24: edition 2. 1828.

MASTERS, M. T. A general view of the genus Pinus. *J. Linn. Soc. Bot.* **35**, 560–659. 1904.

SCHWARZ, O. Über die Systematik und Nomenklatur der europäischen Schwarzkiefern. *Notizbl. Bot. Gart. Berlin* **13**, 226–243. 1936.

SHAW, G. R. The Genus Pinus. *Publ. Arnold Arb.* No. 5. Pp. 96. 1914.

TEUSCHER, H. Bestimmungstabelle für die in Deutschland Klima Kultivierbaren Pinus-Arten. *Mitt. Deutsch. Dendr. Ges.* **1921**, 68–114. 1921.

CUPRESSACEAE

CHAMAECYPARIS

FRANCO, J. do AMARAL. O genero Chamaecyparis Spach. *Agros* **24**, 91–99. 1941.

JUNIPERUS

CUCCHI, C. Indagine geobotanica sui ginepri europei. *Delpinoa* **11**, 171–222. 1958.

PILGER, R. Die Gattung Juniperus. *Mitt. Deutsch. Dendr. Ges.* **43**, 255–269. 1931.

SPACH, E. Révision des Juniperus. *Ann. Sci. Nat. sér.* 2 **16**, 282–305. 1841.

TAXACEAE

TAXUS

PILGER, R. *Taxus,* in Engler, H. G. A. (Ed.), *Das Pflanzenreich* **18** (IV.5), 110–116. 1903.

PILGER, R. Die Taxales. *Mitt. Deutsch. Dendr. Ges.* **25**, 1–28. 1916.

ANGIOSPERMAE

DICOTYLEDONES

RANUNCULACEAE

LOURTEIG, A. Ranunculáceas de Sudamerica templada. *Darwiniana* **9**, 397–608. 1952.

LOURTEIG, A. Ranunculáceas de Sudamerica tropicale. *Mem. Soc. Cienc. Nat. "La Salle"* **16**, 40–42. 1956.

PRANTL, K. Beiträge zur Morphologie und Systematik der Ranunculaceen. *Bot. Jahrb.* **9**, 225–273. 1887.

CALTHA

HUTH, E. Monographie der Gattung Caltha. *Abh. Vortr. Ges. Naturw.* (*Berlin*) **4**. Pp. 33. 1891. Reprinted in *Helios* **9**, 55–78. 1892: *loc. cit.* **9**, 99–103. 1893.

LEONCINI, M. L. Biotipi cariologici e sistematici di Caltha in Italia. *Caryologia* **3**, 336–350. 1951.

MAUGINI, E. Ricerche cito-sistematiche sul genere Caltha in Italia. *Caryologia* **9**, 408–435. 1957.

PANIGRAHI, G. *Caltha in the British flora*, in Lousley, J. E. (Ed.), *Species Studies in the British Flora*, 107–110. London. 1955.

TROLLIUS

HUTH, E. Revision der Arten von Trollius. *Helios* **9**, 1–18. 1891.

SCHIPCZINSKY, N. V. Ueber die geographische Verbreitung und den genetische Zusammentang den Arten der Gattung Trollius. *Bull. Jard. Bot. URSS* **28**, 55–74. 1924.

HELLEBORUS

MERXMÜLLER, H. and PODLECH, D. Über die europäischen Vertreter von Helleborus sect. Helleborus. *Fedde. Rep.* **64**, 2–11. 1961.

SCHIFFNER, V. Die Gattung Helleborus. Eine monografische Skizze. *Bot. Jahrb.* **11**, 97–122. 1889.

SCHIFFNER, V. Monographia Helleborum. *Nova Acta Acad. Leop.-Carol.* **56** (1). Pp. 198. 1890.

ULBRICH, E. Die Arten der Gattung Helleborus (Tourn.) L. *Blätt. Staudenk.* **1938**. Pp. 18. 1938.

ERANTHIS

HUTH, E. Revision der kleinen Ranunculaceen-Gattungen Myosurus, Trautvetteria, Hamadryas, Glaucidium, Hydrastis, Eranthis, Coptis, Anemonopsis, Actaea, Cimicifuga und Xanthorrhiza. *Bot. Jahrb.* **16**, 278–324. 1893.

ACONITUM

GÁYER, G. Vorarbeiten zu einer Monographie der europäischen Aconitum-Arten. *Magyar Bot. Lap.* **5**, 122–137. 1906: *loc. cit.* **6**, 286–303. 1907: *loc. cit.* **8**, 114–206, 310–333. 1909: *loc. cit.* **10**, 194–208. 1911.

GÁYER, G. Nachträge und Berichtigungen zur Bearbeitung der Gattung Aconitum in der Ascherson-Graebnerschen Synopsis. *Magyar Bot. Lap.* **29**, 39–48. 1930.

KOELLE, J. L. C. *Spicilegium Observationum de Aconito.* Erlangae. 1786.

MUNZ, P. A. The cultivated Aconites. *Gentes Herb.* **6**, 463–506. 1945.

Rapaics, R. Systema Aconiti generis. *Növ. Közlem.* **6**, 137–176. 1907.
Regel, E. Conspectus Specierum Generis Aconiti quae in Flora Rossica et in regionibus adjacentibus inveniuntur. *Ann. Sci. Nat. Sér.* 4 *Bot.* **16**, 144–153. 1861.
Reichenbach, H. G. L. *Übersicht der Gattung Aconitum. Grundzüge einer Monographie derselben.* Regensberg. 1819.
Reichenbach, H. G. L. *Monographia Generis Aconiti.* Fasc. 1–2. Pp. iv + 72 plates + 1–6. Lipsiae. 1820: Fasc. 3–4. Pp. 73–100, tt. 7–18. 1821.
Reichenbach, H. G. L. *Illustratio Specierum Aconiti Generis.* Pp. 148. Lipsiae. 1823–27.
Seringe, N. C. Esquisse d'une Monographie du genre Aconitum. *Bull. Soc. Phys. Genève* **1822**, 115–175. 1822.
Stapf, O. Aconitum anglicum. *Bot. Mag.* **1926** (t. 9088). 1926.

DELPHINIUM

Ewan, J. A synopsis of the North American species of Delphinium. *Univ. Colorado Stud. D.* **2**, 55–244. 1945.
Huth, E. Monographie der Gattung Delphinium. *Bot. Jahrb.* **20**, 322–499. 1895.
Pawlowski, B. Studien über mitteleuropäischen Delphinien aus der sogenannten Sektion Elatopsis, I-V. *Bull. Acad. Polon. Sci. Lett., Sér. B,* **1933**, 29–44, 67–81, 91–108, 149–181. 1934.
Pawlowski, B. Dispositio systematica specierum europaearum generis Delphinium L. *Fragm. Fl. Geobot.* **9**, 429–446. 1963.

CONSOLIDA (DELPHINIUM, p.p.)

Huth, E. Monographie der Gattung Delphinium. *Bot. Jahrb.* **20**, 322–499. 1895.
Hylander, N. Några anmarkärkninger om de som prydnadsväxter odlade arterna av Delphinium undersläktet Consolida. *Bot. Not.* **1945**, 75–80. 1945.
M'Nab, W. R. On the British species of Delphinium. *Trans. Proc. Bot. Soc. Edinb.* **9**, 332–336. 1868.
Soó, R. von. Über die mitteleuropäischen Arten und Formen der Gattung Consolida (DC.) S. F. Gray. *Österr. Bot. Zeitschr.* **71**, 233–246. 1922.

NIGELLA

Brand, A. Monographia der Gattung Nigella. *Helios* **13**, 8–38. 1896.
Spenner, F. C. L. *Monographia Generis Nigellae dissertatio inauguralis Botanico-Medico.* Pp. 12. Friburgi. 1829.
Terracciano, A. Revisione monografica delle specie del genere Nigella. *Boll. Orto Bot. Palermo* **1**, 122–153: *loc. cit.* **2**, 19–42. 1898.

ACTAEA

Huth, E. Revision der kleineren Ranunculaceen-Gattungen Myosurus, Trautvetteria, Hamadryas, Glaucidium, Hydrastis, Eranthis, Coptis, Anemonopsis, Actaea, Cimicifuga und Xanthorrhiza. *Bot. Jahrb.* **16**, 278–324. 1893.

ANEMONE

Pritzel, G. A. Anemonarum revisio. *Linnaea* **15**, 561–698. 1841.

Ulbrich, E. Über die systematische Gliederung und geographische Verbreitung der Gattung Anemone L. *Bot. Jahrb.* **37**, 172–256. 1905: *loc. cit.* **37**, 257–334. 1906: (Vortrag). *Verh. Bot. Ver. Brandenb.* **48**, 1–38. 1906.

PULSATILLA

Aichele, D. and Schwegler, H.-W. Die Taxonomie der Gattung Pulsatilla. *Fedde. Rep.* **60**, 1–230. 1957.

Zimmerman, W. Zur Taxonomie der Gattung Pulsatilla Miller. *Fedde. Rep.* **61**, 94–100. 1958.

CLEMATIS

Jouin, E. Die in Deutschland kultivierten, winterharten Clematis. *Mitt. Deutsch. Dendr. Ges.* **1907**, 228–238. 1907.

Kuntze, O. Monographie der Gattung Clematis. *Verh. Bot. Ver. Brandenburg* **26**, 83–202. 1885.

Kuntze, O. Nachtrage zur Clematis-Monographie. *Verh. Zool.-Bot. Ges. Wien* **37**, 47–50. 1887.

RANUNCULUS

Baillon, H. E. *Monographie des Renonculacées.* Pp. 88. Paris. 1867.

Benson, L. A treatise on the North American Ranunculi. *Amer. Midl. Nat.* **40**, 1–261. 1948.

Benson, L. Supplement to a treatise on the North American Ranunculi. *Amer. Midl. Nat.* **52**, 328–369. 1954.

Davis, K. C. Native and cultivated Ranunculi of North America and segregated genera. *Minn. Bot. Stud.* **2**, 459–507. 1900.

Freyn, J. Zur Kenntnis einiger Arten der Gattung Ranunculus. *Bot. Centralbl.* **6** (26). Pp. 36. 1881.

Freyn, J. Beiträge zur Kenntnis einiger Arten der Gattung Ranunculus. *Bot. Centralbl.* **41**, 1–6, 33–37, 73–78, 129–134. 1890.

Gray, A. A revision of Ranunculus in North America. *Proc. Amer. Acad. Arts Sci.* **21**, 363–378. 1886.

Pons, G. Saggio di una rivista critica delle specie Itáliene del genere Ranunculus L. *Nuov. Giorn. Bot. Ital., nuov. ser.* **5**, 210–254, 353–392. 1898: *loc. cit.* **8**, 5–27. 1901.

Sect. ***AURICOMUS***

(See also *Ranunculus* (general))

CEDERCREUTZ, C. Einige neue Sippen der Ranunculus auricomus-Gruppe. *Acta Soc. Fauna Fl. Fenn.* **78** (4), 1–7 + 10 plates. 1965.

FAGERSTROM, L. Neue Sippen des Ranunculus auricomus-Komplexes aus Finnland. *Acta Soc. Fauna Fl. Fenn.* **78** (1), 1–10 + 5 plates. 1965.

HAAS, P. A. Neue Süddeutsche Arten aus dem Formenkreis des Ranunculus auricomus. *Ber. Bayer. Bot. Ges.* **29**, 5–12. 1952.

HAAS, P. A. Neuer Beitrag zur Kenntnis des Formenkreises von Ranunculus auricomus L. in Süddeutschland. *Ber. Bayer. Bot. Ges.* **30**, 27–32. 1954.

JASIEWICZ, A. De Ranunculus e circulo Auricomi Owcz. in regione Cracoviensi nec non in Carpatorum parte boreali crescentibus. *Fragm. Fl. Geobot.* **2**, 62–110. 1956.

JULIN, E. Der Formenkreis des Ranunculus auricomus L. in Schweden, 1. Diagnosen und Fundortsangaben einige sippen aus Södermanland. *Arkiv Bot. ser.* 2 **6**, 1–108. 1965.

KOCH, W. Schweizerische Arten aus der Verwandtschaft des Ranunculus auricomus L. *Ber. Schweiz. Bot. Ges.* **42**, 740–753. 1933.

KOCH, W. Zweiter Beitrag zur Kenntnis des Formenkreises von Ranunculus auricomus L. *Ber. Schweiz. Bot. Ges.* **49**, 541–554. 1939.

MARKLUND, G. Der Ranunculus auricomus-Komplex in Finnland, 1. *Flora Fennica* **3**, 1–128 + 94 plates. 1961: 2, *loc. cit.* **4**, 1–104 + 94 plates. 1965.

NYÁRÁDY, E. I. Despre grupa "Auricomus" a genului Ranunculus. *Bul. Grăd. Bot. Cluj* **13**, 86–101. 1933.

ROUSI, A. Cytotaxonomy and reproduction in the apomictic Ranunculus auricomus group. *Ann. Bot. Soc. Zool.-Bot. Fenn. Vanamo* **29** (2), 1–64. 1956.

ROZANOVA, M. Versuch einer analytischen Monographie der Conspecties Ranunculus auricomus Korsh. *Trav. Inst. Sci. Nat. Peterhoff* **1932**, No. 8. 1932.

*SCHILLER, Z. *Math. Term. Értesitö* **35**, 361–447. 1917.

SCHWARZ. O. Beiträge zur Kenntnis kritischer Formenkreise im Gebiete der Flora von Thüringen, iv. Ranunculus ser. Auricomi. *Mitt. Thür. Bot. Ges.* **1**, 120–143. 1949.

SOÓ, R. von. Die Ranunculus auricomus L. emend Korsh. Artengruppe in der Flora Ungarns und der Karpaten. *Acta Bot. Hung.* **10**, 221–237. 1964: *loc. cit.* **11**, 395–404. 1965.

Sect. ***FLAMMULA***

(See also *Ranunculus* (General))

GLÜCK, H. *Biologische und Morphologische Untersuchungen über Wasser- und Sumpfgewächse* **4**. Pp. viii + 746 + Taf. 8. Jena. 1924.

PADMORE, P. A. The varieties of Ranunculus flammula L. and the status of R. scoticus E. S. Marshall and of R. reptans. L. *Watsonia* **4**, 19–27. 1957.

Subgen. *BATRACHIUM*

(See also *Ranunculus* (General))

Babington, C. C. On the Batrachian Ranunculi of Britain. *Ann. Mag. Nat. Hist. ser.* 2 **16**, 385–404. 1855. Reprinted in *Trans. Proc. Bot. Soc. Edinb.* **5**, 65–84. 1858.

Butcher, R. W. Notes on water buttercups. *Nat.* **1960**, 123–125. 1960.

Cook, C. D. K. Studies in Ranunculus subgenus Batrachium (DC.) A. Gray, 2. General morphological considerations in the taxonomy of the subgenus. *Watsonia* **5**, 294–303. 1963.

Cook, C. D. K. A monographic study of Ranunculus subgenus Batrachium (DC.) A. Gray. *Mitt. Bot. Staats. München* **6**, 47–237. 1966.

Drew, W. B. The North American representatives of Ranunculus § Batrachium. *Rhodora* **38**, 1–47, 1936.

Dumortier, B. C. Monographie du genre Batrachium. *Bull. Soc. Bot. Belg.* **2**, 207–219. 1863.

Félix, M. A. Études monographiques sur les Renoncules françaises de la Section Batrachium. *Bull. Soc. Bot. France* **57**, 406–412 + xxxiv–xl. 1910: *loc. cit.* **58**, 97–103. 1911: *loc. cit.* **59**, 112–120 + lxi–lxvi. 1912: *loc. cit.* **60**, 258–266. 1913: *loc. cit.* **61**, 107–112, 352–355. 1914: *loc. cit.* **63**, 56–66. 1916: *loc. cit.* **72**, 774–778. 1925: *loc. cit.* **73**, 77–86. 1926: *loc. cit.* **74**, 277–280. 1927.

Gelert, O. Studier over Slaegten Batrachium. *Bot. Tidsskr.* **19**, 7–35. 1894.

Glück, H. *Biologische und Morphologische Untersuchungen über Wasser- und Sumpfgewächse* **4**, 136–258. Jena. 1924.

Godron, C. Essai sur les renoncules a fruits ridées transversalement. *Mém. Soc. Roy. Sci. Lett. Arts Nancy* **1839**, 41–75. 1839.

Hiern, W. P. On the forms and distribution over the world of the Batrachium Section of Ranunculus. *J. Bot.* **9**, 43–49, 65–72, 97–101. 1871.

Lawson, G. Revision of Batrachium in Canada. *Trans. Roy. Soc. Canada* **2**: sect. 4, 44–46. 1884.

Meikle, R. D. The Batrachian Ranunculi of the Orient. *Notes Roy. Bot. Gard. Edinb.* **23**, 13–22. 1959.

Pearsall, W. H. Notes on the British Batrachia. *Rep. Bot. Soc. & E.C.* **6**, 440–452. 1921.

Pearsall, W. H. The British Batrachia. *Rep. Bot. Soc. & E.C.* **8**, 811–837. 1928.

Rossman, J. *Beiträge zur Kenntniss der Wasserhahnenfüsse Ranunculus sect. Batrachium.* Pp. 62. Giessen. 1854.

Tullberg, A. S. Öfversigt af de skandinaviska arterna af slägted Ranunculus L., gruppen Batrachium DC. *Bot. Not.* **1873**, 65–71. 1873.

Williams, F. N. Critical study of Ranunculus aquatilis L. *J. Bot.* **46**, 11–22, 44–52. 1908.

ADONIS

HUTH, E. Revision der Arten von Adonis und Knowltonia. *Helios* **8**, 61–73.1890.

RIEDL, H. Revision der einjährigen Arten von Adonis L. *Ann. Nat. Mus.* (*Wien*) **66**, 51–90.1963.

MYOSURUS

CAMPBELL, G. R. The genus Myosurus L. (Ranunculaceae) in North America. *Aliso* **2**, 398–403.1952.

HUTH, E. Revision der kleineren Ranunculaceen—Gattungen Myosurus, Trautvetteria, Hamadryas, Glaucidium, Hydrastis, Eranthis, Coptis, Anemonopsis, Actaea, Cimicifuga and Xanthorrhiza. *Bot. Jahrb.* **16**, 278–324.1893.

AQUILEGIA

BAKER, J. G. A synopsis of the known forms of Aquilegia. *Gard. Chron. new ser.* **10**, 19–20, 76, 111, 203.1878.

BECHERER, A. Bemerkungen zur Gattung Aquilegia. *Ber. Schweiz. Bot. Ges.* **68**, 289–294.1958.

MUNZ, P. A. Aquilegia, the cultivated and wild columbines. *Gentes Herb.* **7**, 1–150.1946.

PAYSON, E. P. The North American species of Aquilegia. *Contr. U.S. Nat. Herb.* **20**, 133–157.1918.

RAPAICS, R. Die genere Aquilegia. *Bot. Közl.* **8**, 117–136.1909.

ZIMMETER, A. Verwandtschafts-Verhältnisse und geographische Verbreitung der in Europa einheimischen Arten der Gattung Aquilegia. *Jahresb. Staats-ober-Realschule Steyr* **5**, 1–66.1875.

THALICTRUM

BOIVIN, B. American Thalictra and their Old World allies. *Rhodora* **46**, 337–377, 391–445, 453–487.1944.

LECOYER, J.-C. Monographie du genre Thalictrum. *Bull. Soc. Bot. Belg.* **24**, 78–324 + 45 plates.1885.

PAEONIACEAE

PAEONIA

ANDERSON, G. A monograph of the genus Paeonia. *Trans. Linn. Soc.* **12**, 248–290.1818.

CULLEN, J. and HEYWOOD, V. H. Notes on the European species of Paeonia. *Fedde. Rep.* **69**, 32–35.1964.

HUTH, E. Monographie der Gattung Paeonia. *Bot. Jahrb.* **14**, 258–276.1891.

LYNCH, R. I. A new classification of the genus Paeonia. *J. Roy. Hort. Soc., new ser.* **12**, 428–445.1890.

STERN, F. C. *A Study of the Genus Paeonia.* Pp. viii + 155. London. 1946.

BERBERIDACEAE

EPIMEDIUM

STEARN, W. T. Epimedium and Vancouveria (Berberidaceae), a monograph. *J. Linn. Soc. Bot.* **51**, 409–535. 1938.

BERBERIS

AHRENDT, L. W. A. A survey of the genus Berberis L. in Asia. *Supplement J. Bot.* **80**, Pp. 116. 1941–42.

AHRENDT, L. W. A. Berberis and Mahonia. A taxonomic revision. *J. Linn. Soc. Bot.* **57**, 1–410. 1961.

BAUER, G. Beiträge zur Kenntnis der Berberidaceen. *Mitt. Deutsch. Dendr. Ges.* **1932**, 42–46. 1932.

LANGE, J. *Løvfaeldende Berberis.* Pp. 72. København. 1961.

SCHNEIDER, C. K. Die Gattung Berberis (Euberberis) Vorarbeiten für eine Monographie. *Bull. Herb. Boiss., sér.* 2 **5**, 33–48, 133–148, 391–403, 449–464, 655–670, 800–831. 1905.

SCHNEIDER, C. K. Die Gattung Berberis (Euberberis). *Mitt. Deutsch. Dendr. Ges.* **1905**, 111–124. 1905.

SCHNEIDER, C. K. Bemerkungen über die Berberis des Herbar Schrader. *Mitt. Deutsch. Dendr. Ges.* **1906**, 173–181. 1906.

SCHNEIDER, C. K. Weitere Beiträge zur Kenntnis der Gattung Berberis (Euberberis). *Bull. Herb. Boiss., sér.* 2 **8**, 192–204, 258–266. 1908.

USTERI, A. Das Geschlecht der Berberitzen. *Mitt. Deutsch. Dendr. Ges.* **1899**, 77–95. 1899.

MAHONIA

AHRENDT, L. W. A. Berberis and Mahonia. A taxonomic revision. *J. Linn. Soc. Bot.* **57**, 1–410. 1961.

FEDDE, F. Versuch einer Monographie der Gattung Mahonia. *Bot. Jahrb.* **31**, 30–133. 1901.

JOUIN, E. Die in Lothringen winterharten Mahonien. *Mitt. Deutsch. Dendr. Ges.* **1910**, 86–91. 1910.

TAKEDA, H. Contributions to the knowledge of the Old World species of the genus Mahonia. *Notes Roy. Bot. Gard. Edinb.* **6**, 209–245. 1917.

NYMPHAEACEAE

NYMPHAEA

CONARD, S. *The Waterlilies. A Monograph of the Genus Nymphaea.* Pp. 279. Washington. 1905.

HENKEL, F., REHNELT, F. and DITTMANN, L. *Das Buch der Nymphaeaceen oder Seerosengewächse.* Pp. 158. Darmstadt. 1907.

HESLOP-HARRISON, Y. Nymphaea L. em. Sm. (Biological Flora). *J. Ecol.* **43**, 719–734. 1955.

SCHUSTER, J. Zur Systematik von Castalia und Nymphaea. *Bull. Herb. Boiss., ser.* 2 **7**, 858–868, 901–916, 981–996. 1907: *loc. cit.* **8**, 65–74. 1908.

NUPHAR

Beal, O. E. Taxonomic revision of the genus Nuphar Sm. of North America and Europe. *J. Elisha Mitchell Sci. Soc.* **72**, 319–346. 1956.

Heslop-Harrison, Y. Nuphar Sm. (Biological Flora). *J. Ecol.* **43**, 342–364. 1955.

CERATOPHYLLACEAE

CERATOPHYLLUM

Glück, H. *Biologische und Morphologische Untersuchungen über Wasser- und Sumpfgewächse* 2. Pp. xvii + 256 + Taf. 6. Jena. 1906.

Pearl, R. Variation and differentiation in Ceratophyllum. *Carnegie Inst. Washington Publ.* **58**. 1907.

Sandwith, C. The Honeworts and their occurrence in Britain. *Ann. Rep. Proc. Bristol Nat. Soc.* **6**, 303–311. 1926.

PAPAVERACEAE

PAPAVER

Baillon, H. E. *Monographie des Papavéracées et des Capparidacées*. Paris. 1871.

Elkan, L. *Tentamen Monographie Generis Papaver*. Konigsberg. 1837.

Fedde, F. *Papaver*, in Engler, H. G. A. (Ed.), *Das Pflanzenreich* 40 (IV. 104), 288–386. 1909.

Markgraf, F. *Papaver*, in Hegi, G. *Illustr. Fl. Mitteleur.* Edition 2, **4** (1), 29–49. 1958.

MECONOPSIS

Fedde, F. *Meconopsis*, in Engler, H. G. A. (Ed.), *Das Pflanzenreich* **40** (IV. 104), 247–271. 1909.

Prain, D. Some additional species of Meconopsis. *Kew Bull.* **1915**, 129–177. 1915.

Taylor, G. *An Account of the Genus Meconopsis*. Pp. xiii + 130. London. 1934.

ARGEMONE

Ownbey, G. B. Monograph of the genus Argemone for North America and the West Indies. *Mem. Torrey Bot. Club* **21** (1), 1–149. 1958.

ROEMERIA

Fedde, F. *Roemeria*, in Engler, H. G. A. (Ed.), *Das Pflanzenreich* **40** (IV. 104), 238–244. 1909.

GLAUCIUM

Fedde, F. *Glaucium*, in Engler, H. G. A. (Ed.), *Das Pflanzenreich* **40** (IV. 104), 221–238. 1909.

Turrill, W. B. A study of variation in Glaucium flavum. *Kew Bull.* **1933**, 174–184. 1933.

CHELIDONIUM

FEDDE, F. *Chelidonium*, in Engler, H. G. A. (Ed.), *Das Pflanzenreich* **40** (IV. 104), 212–215. 1909.

PRAIN, D. A revision of the genus Chelidonium. *Bull. Herb. Boiss.* **3**, 570–587. 1895.

ESCHSCHOLZIA

FEDDE, F. *Eschscholzia*, in Engler, H. G. A. (Ed.), *Das Pflanzenreich* **40** (IV. 104), 144–202. 1909.

FUMARIACEAE

DICENTRA

STERN, K. R. Revision of Dicentra (Fumariaceae). *Brittonia* **13**, 1–57. 1961.

CORYDALIS

FEDDE, F. Beiträge zur Kenntnis der europäischen Arten der Gattung Corydalis. *Fedde. Rep.* **16**, 49–60, 187–192. 1919.

POELLNITZ, K. von. Zur Kenntnis von Corydalis § Pesga llinaceus Irmisch. *Fedde. Rep.* **44**, 154–157. 1938: *loc. cit.* **45**, 96–112. 1938.

RYBERG, M. A taxonomical survey of the genus Corydalis Ventenat with reference to cultivated species. *Acta Hort. Berg.* **17**, 115–175. 1955.

FUMARIA

BABINGTON, C. C. On the British species of Fumaria. *Trans. Proc. Bot. Soc. Edinb.* **1**, 31–38. 1841.

HAMMAR, O. N. *En Monografi öfver slägtet Fumaria.* Pp. xvi + 58. Lund. 1854.

HAMMAR, O. N. *Monographia Generis Fumariarum.* Uppsala. 1857. Reprinted in *Nova Acta Regiae Soc. Sci. Upsal.-Kong. Vet. Soc. sér.* 3 **2** (8). 1858.

HAUSSKNECHT, C. von. Beitrag zur Kenntnis der Arten von Fumaria sect. Sphaerocapnos DC. *Flora* **56**, 401–414, 417–425, 441–446, 456–462, 485–496, 505–526, 536–544, 546–560, 562–568. 1873.

PARLATORE, F. Monografia delle Fumariée. *Giorn. Bot. Ital.* **1**, 124–160. 1844.

PUGSLEY, H. W. The genus Fumaria in Britain. *Supplement J. Bot.* **50**. Pp. 76. 1912.

PUGSLEY, H. W. A revision of the genera Fumaria and Rupicapnos. *J. Linn. Soc. Bot.* **44**, 233–355. 1919: *loc. cit.* **47**, 427–469. 1927: *loc. cit.* **49**, 93–113. 1932: *loc. cit.* **49**, 517–529. 1934: *loc. cit.* **50**, 541–559. 1937.

SELL, P. D. Taxonomic and nomenclatural notes on European Fumaria species. *Fedde. Rep.* **68**, 174–178. 1963.

CRUCIFERAE

JANCHEN, E. Das System der Cruciferen. *Österr. Bot. Zeitschr.* **91**, 1–28. 1942.

BRASSICA

BAILEY, L. H. The cultivated Brassicas. *Gentes Herb.* **1**, 53–108. 1922.

BAILEY, L. H. Certain noteworthy Brassicas. *Gentes Herb.* **4**, 319–330. 1940.

GATES, R. R. Genetics and taxonomy of the cultivated Brassicas and their wild relatives. *Bull. Torrey Bot. Club* **77**, 19–28. 1950.

HELM, J. Morphologisch-taxonomische Gliederung der Kultursippen von Brassica oleracea L. *Kulturpflanze* **11**, 92–210. 1963.

ONNO, M. Die Wildformen aus dem Verwandtschaftskreis "Brassica oleraea L." *Österr. Bot. Zeitschr.* **82**, 309–334. 1933.

SCHULZ, O. E. *Brassica*, in Engler, H. G. A. (Ed.), *Das Pflanzenreich* **70** (IV. 105), 21–84. 1919.

SUN, V. G. The evaluation of taxonomic characters of cultivated Brassica with a key to species and varieties. *Bull. Torrey Bot. Club* **73**, 244–281, 370–377. 1946.

ERUCASTRUM

SCHULZ, O. E. *Erucastrum*, in Engler, H. G. A. (Ed.), *Das Pflanzenreich* **70** (IV. 105), 88–106. 1919.

RHYNCHOSINAPIS

SCHULZ, O. E. *Rhynchosinapis*, in Engler, H. G. A. (Ed.), *Das Pflanzenreich* **70** (IV. 105), 106–116. 1919.

SINAPIS

SCHULZ, O. E. *Sinapis*, in Engler, H. G. A. (Ed.), *Das Pflanzenreich* **70** (IV. 105), 118–136. 1919.

HIRSCHFELDIA

SCHULZ, O. E. *Hirschfeldia*, in Engler, H. G. A. (Ed.), *Das Pflanzenreich* **70** (IV. 105), 136–143. 1919.

DIPLOTAXIS

LUBBERT, G. Vergleichende cytologische, morphologische und physiologische Untersuchungen innerhalb der Gattung Diplotaxis. *Beitr. Biol. Pflanzen* **28**, 254–293. 1951.

NÈGRE, R. Les Diplotaxis du Maroc, de l'Algérie et de la Tunisie. *Mém. Soc. Sci. Nat. Phys. Maroc, Bot. nouv. sér.* **1**, 1–118. 1960.

ROUY, G. Étude des Diplotaxis européens de la section Brassicaria. *Rev. Sci. Nat.* **1882**, 423–436. 1882.

SCHULZ, O. E. *Diplotaxis*, in Engler, H. G. A. (Ed.), *Das Pflanzenreich* **70** (IV. 105), 149–180. 1919.

ERUCA

SCHULZ, O. E. *Eruca,* in Engler, H. G. A. (Ed.), *Das Pflanzenreich* **70** (IV. 105), 180–190. 1919.

RAPHANUS

SCHULZ, O. E. *Raphanus,* in Engler, H. G. A. (Ed.), *Das Pflanzenreich* **70** (IV. 105), 194–210. 1919.

THELLUNG, A. *Raphanus,* in Hegi, G. *Ill. Fl. Mitteleur.* **4** (1), 272–286. 1918.

CRAMBE

SCHULZ, O. E. *Crambe,* in Engler, H. G. A. (Ed.), *Das Pflanzenreich* **70** (IV. 105), 228–249. 1919.

RAPISTRUM

SCHULZ, O. E. *Rapistrum,* in Engler, H. G. A. (Ed.), *Das Pflanzenreich* **70** (IV. 105), 252–261. 1919.

CAKILE

BALL, P. W. A revision of Cakile in Europe. *Fedde. Rep.* **69**, 35–40. 1964.

POBEDIMOVA, E. Notae de genere Cakile Mill. *Not. Syst. (Leningrad)* **15**, 61–77. 1953.

POBEDIMOVA, E. Genus Cakile Mill. (Pars Specialis). *Nov. Syst. Plant. Vasc.* **1964**, 90–128. 1964.

SCHULZ, O. E. *Cakile,* in Engler, H. G. A. (Ed.), *Das Pflanzenreich* **84** (IV. 105 (2)), 18–28. 1923.

CONRINGIA

SCHULZ, O. E. *Conringia,* in Engler, H. G. A. (Ed.), *Das Pflanzenreich* **84** (IV. 105 (2)), 84–94. 1923.

BISCUTELLA

GUINEA, E. El Genero Biscutella. *Anal. Inst. Bot. Cav.* **21**, 387–405. 1963.

MACHATSCHKI-LAURICH, B. Die Arten der Gattung Biscutella L. sectio Thlaspidium (Med.) DC. *Bot. Archiv.* **13**, 1–115. 1926.

MALINOWSKI, E. Monographie du genre Biscutella. *Bull. Acad. Sci. Cracov. Sér. B* **1910**, 111–139. 1910.

LEPIDIUM

HITCHCOCK, C. L. The genus Lepidium in the United States. *Madroño* **3**, 265–320. 1936.

HITCHCOCK, C. L. The South American species of Lepidium. *Lilloa* **11**, 75–134. 1945.

MULLIGAN, G. The genus Lepidium in Canada. *Madroño* **16**, 77–90. 1961.

THELLUNG, A. Die Gattung Lepidium (L.) R. Br. Eine monographische Studie. *Neue Denkschr. Allg. Schweiz. Ges. Naturw.* **41**, abh. 1–340. 1906.

CORONOPUS

MUSCHLER, R. Die Gattung Coronopus (L.) *Bot. Jahrb.* **41**, 111–147. 1908.

CARDARIA

MULLIGAN, G. A. and FRANKTON, C. Taxonomy of the genus Cardaria with particular reference to the species introduced into North America. *Canad. J. Bot.* **40**, 1411–1425. 1962.

THLASPI

PAYSON, E. B. The genus Thlaspi in North America. *Wyoming Univ. Publ. Bot.* **1** (6), 145–163. 1926.

THELLUNG, A. *Thlaspi*, in Hegi, G., *Illustr. Fl. Mitteleur.* **4** (1), 116–134. 1914.

CAPSELLA

ALMQUIST, E. Studien über die Capsella bursa-pastoris (L.) *Acta Hort. Berg.* **4** (6), 3–91. 1907: *loc. cit.* **7** (2), 41–95. 1923.

ALMQUIST, E. Bursa pastoris Weber. *Supplement Rep. Bot. Soc. & E.C.* **6**, 179–207. 1921.

COCHLEARIA

COWAN, M'T. A revision of the genus Cochlearia L. in Britain, 1. Cochlearia danica Linn. *Trans. Proc. Bot. Soc. Edinb.* **26**, 136–140. 1913.

LOVKIST, B. Nagot om de skånska Cochlearia-arterna. *Bot. Not.* **116**, 326–330. 1962.

SAUNTE, L. H. Cyto-genetical studies in the Cochlearia officinalis complex. *Hereditas* **41**, 499–515. 1955.

SUBULARIA

HILTNER, L. Untersuchungen über die Gattung Subularia. *Bot. Jahrb.* **7**, 264–272. 1886.

MULLIGAN, G. A. and CALDER, J. A. The genus Subularia (Cruciferae). *Rhodora* **66**, 127–135. 1964.

ALYSSUM

BAUMGARTNER, J. Die ausdauernden Arten der sectio Eualyssum aus der Gattung Alyssum. *Jahresb. Landes-Lehresem. Wiener-Neustadt* **34**, xvi + 1–35. 1907: *loc. cit.* **35**, 3–57. 1908: *loc. cit.* **36**, 3–32. 1909: *Verh. Kaiser Franz Josef Landes-Gymnasium Baden bei Wien* **48**, 3–18. 1911.

NYÁRÁDY, E. I. Studiu preliminar asupra unor specii de Alyssum din sectia Odontarrhena. *Bul. Grăd. Bot. Cluj* **7**, 1–51, 65–160. 1927: *loc. cit.* **8**, 152–156. 1928: *loc. cit.* **9**, 1–68. 1929.

NYÁRÁDY, E. I. Synopsis specierum, variationum et formarum Sectionis Odontarrhenae generis Alyssum. *An Acad. Rep. Pop. Române* Ser. A **1** (3), 67–200. 1949.

DRABA

BALDACCI, A. Monografia della sezione "Aizopsis DC." del genere Draba L. *Nuov. Giorn. Bot. Ital. nuov. ser.* **1**, 103–121. 1894.

FERNALD, M. L. Draba in temperate northeastern America. *Rhodora* **63**, 241–260, 285–304, 314–344, 353–372, 392–404. 1934.

SCHULZ, O. E. *Draba*, in Engler, H. G. A. (Ed.), *Das Pflanzenreich* **89** (IV. 105), 16–343. 1927.

WEINGERL, H. Beiträge zu einer Monographie der europäisch-asiatischen Arten aus der Gattung Draba sect. Leucodraba. *Bot. Arch.* **4**, 9–109. 1923.

EROPHILA

DRUCE, G. C. The British Erophila. *Rep. Bot. Soc. & E.C.* **9**, 177–198. 1930.

MARANNE, I. Les Erophila DC. *Bull. Soc. Bot. France* **60**, 276–281, 345–353, 379–389, 422–425. 1913.

MATUSZKIEWICZ, W. Taxonomic researches on Erophila verna DC. *Ann. Univ. Mariae-Curie* **3**, 19–47. 1948.

SCHULZ, O. E. *Erophila*, in Engler, H. G. A. (Ed.), *Das Pflanzenreich* **89** (IV. 105 (3)), 343–374. 1927.

WINGE, Ö. E. Taxonomic and evolutionary studies in Erophila, based on cytogenetic investigations. *Compt. Rend. Trav. Lab. Carlsb.* (*Sér. Physiol.*) **25**, 41–74. 1940.

CARDAMINE

BANACH-POGAN, E. Dalsze badania cytologiczne nad gatunkami rodzaju Cardamine L. *Acta Soc. Bot. Pol.* **24**, 275–286. 1955.

HUSSEIN, F. Chromosome races in Cardamine pratensis in the British Isles. *Watsonia* **3**, 170–174. 1955.

LÖVKIST, B. The Cardamine pratensis complex—outlines of its cytogenetics and taxonomy. *Symb. Bot. Upsal.* **14** (2), 1–131. 1956.

LÖVKIST, B. De skandinaviska arterna i Cardamine pratensis–komplexet. *Bot. Not.* **110**, 237–250. 1957.

SCHULZ, O. E. Monographie der Gattung Cardamine. *Bot. Jahrb.* **32**, 280–623. 1903.

***NASTURTIUM* (*RORIPPA* p.p.)**

HOWARD, H. W. and LYON, A. G. The identification and distribution of the British watercress species. *Watsonia* **1**, 228–233. 1950.

BARBAREA

LANGE, T. Sveriges Barbarea-arter. *Bot. Not.* **1937**, 216–230. 1937.

ARABIS (includes ***TURRITIS***)

Hopkins, M. Arabis in eastern and central North America. *Rhodora* **39**, 63–98, 106–148, 155–186. 1937

Klášterský, I. and Novotná, I. Komplex Arabis hirsuta. *Preslia* **34**, 387–393. 1962.

Rollins, R. C. A monographic study of Arabis in western North America. *Rhodora* **43**, 289–325, 348–411, 425–481. 1941.

RORIPPA

Borbás, V. Floristicai adatok különös tekintettel a Roripákra. *Értek. Term. Köréb. Magyar Tud. Akad.* **9**, 3–64. 1879.

AUBRIETA

Mattfeld, J. The species of the genus Aubrieta Adanson. *Quart. Bull. Alpine Gard. Soc.* **7**, 157–181, 217–227. 1939. Translated into English by permission of the author from *Blätt. Staudenk.* **1** (1–7) (1937), by V. Higgins and W. T. Stearn.

MATTHIOLA

Britten, J. The genus Matthiola in Britain. *J. Bot.* **38**, 168–169. 1900.

Conti, P. Classification et distribution des espèces Européennes du genre Matthiola. *Bull. Herb. Boiss.* **5**, 31–59, 315–325. 1897.

Conti, P. Les espèces du genre Matthiola. *Mém. Herb. Boiss.* **1** (18), 1–86. 1900.

MALCOLMIA

Ball, P. W. A review of Malcolmia maritima and allied species. *Fedde. Rep.* **68**, 179–186. 1963.

HESPERIS

Borbás, V. Hazánk meg a Balkán Hesperis-ei. *Magyar Bot. Lap.* **1**, 161–167, 196–205, 229–237, 261–272, 304–313, 344–348, 369–380. 1902: *loc. cit.* **2**, 12–23. 1903.

Dvořák, F. Ad Hesperidis L. generis species in Haemo crescentes adnotationes aliquae. *Preslia* **38**, 57–66. 1966.

Fournier, E. Monographie du genre Hesperis. *Bull. Soc. Bot. France* **13**, 326–362. 1866.

Tzvelev, N. Genus Hesperis L. in URSS. *Not. Syst.* (*Leningrad*) **19**, 114–155. 1959.

ERYSIMUM

Gay, J. E. *Erysimorum quorundum novorum Diagnoses simulque Erysimi muralis Descriptionem praemittit, Monographiam Generis editurus J. Gay.* Parisiis. 1842.

KONĚTOPSKÝ, A. Nejduležitějšti výsledky taxonomické revise československých druhů rodu Erysimum L. *Preslia* **35**, 135–145. 1963.

ROSSBACH, G. B. The genus Erysimum (Cruciferae) in North America north of Mexico—a key to the species and varieties. *Madroño* **14**, 261–267. 1958.

ALLIARIA

SCHULZ, O. E. *Alliaria*, in Engler, H. G. A. (Ed.), *Das Pflanzenreich* **86** (IV. 105 (2)), 20–26. 1924.

SISYMBRIUM

FOURNIER, P. N. E. *Theses . . . Recherches anatomiques et taxonomiques sur la famille des Crucifères et sur le genre Sisymbrium en particulier.* Pp. 154. Paris. 1865.

PAYSON, E. B. Species of Sisymbrium native to North America north of Mexico. *Wyoming Univ. Publ. Bot.* **1**, 1–27. 1922.

SCHULZ, O. E. *Sisymbrium*, in Engler, H. G. A. (Ed.), *Das Pflanzenreich* **86** (IV. 105(2)), 46–157. 1924.

ARABIDOPSIS

SCHULZ, O. E. *Arabidopis*, in Engler, H. G. A. (Ed.), *Das Pflanzenreich* **86** (IV. 105 (2)), 268–285. 1924.

CAMELINA

TEDIN, O. Vererbung, Variation und Systematik in der Gattung Camelina. *Hereditas* **6**, 275–386. 1925.

*ZINGER, N. *Trav. Mus. Bot. Acad. Pétersb.* **6**, 1–303. 1909.

DESCURAINIA

DETLING, L. E. A revision of the North American species of Descurainia. *Amer. Midl. Nat.* **22**, 481–520. 1939.

SCHULZ, O. E. *Descurainia*, in Engler, H. G. A. (Ed.), *Das Pflanzenreich* **86** (IV. 105 (2)), 305–346. 1924.

RESEDACEAE

RESEDA

MULLER, J. Monographie de la famille des Résédacées. *Neue Denkschr. Allg. Schweiz. Ges. Naturw.* **16**. Pp. 239 + Taf. 10. 1858.

VIOLACEAE

VIOLA

Subgen. ***VIOLA***

BECKER, W. Violenstudien. *Beih. Bot. Centr.* **26**, 1–44. 1909: *loc. cit.* **26**, 289–390. 1910. Reprinted as *Violae Europaeae. Systematische Bearbeitung der Violen Europas und seiner benachbarten Gebeite.* Pp. 153. Dresden. 1910.

FOTHERGILL, P. G. Studies in Viola, 4. The somatic cytology and taxonomy of our British species of the genus Viola. *New Phyt.* **43**, 25–35. 1944.

GADELLA, T. W. J. A cytotaxonomic study of Viola in the Netherlands. *Acta Bot. Neerl.* **12**, 17–39. 1963.

GREGORY, Mrs. E. S. *British Violets.* Pp. xxiii + 108. Cambridge. 1912.

SCHMIDT, A. Zytotaxonomische Untersuchungen an europäischen Viola-Arten der Sektion Nomimium. *Österr. Bot. Zeitschr.* **108**, 20–88. 1961.

TODD, E. E. A short survey of the genus Viola. *J. Roy. Hort. Soc.* **55**, 223–243. 1930.

WITTROCK, V. B. Viola-studier. *Acta Hort. Berg.* **2** (1), 1–142. 1897: *loc. cit.* **2** (7), 1–78. 1895.

Subgen. ***MELANIUM***

BECKER, W. Violenstudien. *Beih. Bot. Centr.* **26**, 1–44. 1909: *loc. cit.* **26**, 289–390. 1910. Reprinted as *Violae Europaeae. Systematische Bearbeitung der Violen Europas und seiner benachbarten Gebiete.* Pp. 153. Dresden. 1910.

CLAUSEN, J. Cyto-genetic and taxonomic investigations on Melanium violets. *Hereditas* **15**, 219–308. 1931.

DRABBLE, E. The British Pansies. *Supplement J. Bot.* **47**. Pp. 32. 1909.

DRABBLE, E. Notes on the British Pansies. *J. Bot.* **64**, 263–271. 1926: *loc. cit.* **65**, 42–53, 161–171, 213–219. 1927: *loc. cit.* **66**, 129–132. 1928.

DRABBLE, E. A key to the British pansies. *J. Bot.* **67**, 69–74. 1929.

GREGORY, Mrs. E. S. *British Violets.* Pp. xxiii + 108. Cambridge. 1912.

WITTROCK, V. B. Viola-studier. *Acta Hort. Berg.* **2** (1), 1–42. 1897: *loc. cit.* **2** (7), 1–78. 1895.

POLYGALACEAE

CHODAT, R. Monographia Polygalacearum. *Mém. Soc. Phys. Genève* 1890. *Supplement* **7**, 1–143 + 12 plates. 1891: *loc. cit.* **31** (2), xii + 1–500 + 23 plates. 1893.

POLYGALA

BABINGTON, C. C. British Polygala. *Trans. Proc. Bot. Soc. Edinb.* **4**, 165–169. 1853.

BENNETT, A. W. A review of the British species and subspecies of Polygala. *J. Bot.* **14**, 168–174. 1877.

CHODAT, R. Révision et critique des Polygala suisses. *Bull. Soc. Bot. Genève* **5**, 123–185. 1889.

CHODAT, R. Monographia Polygalacearum. *Mém. Soc. Phys. Genève* 1890. *Supplement* **7**, 1–143 + 12 plates. 1891: *loc. cit.* **31** (2), xii + 1–500 + 23 plates. 1893.

GUTTIFERAE

CHOISY, D. *Prodromus d'une Monographie de la Famille des Hypéricinées.* Pp. 70. Genève and Paris. 1821.

HYPERICUM

CHOISY, D. *Prodromus d'une Monographie de la Famille des Hypéricinées*. Pp. 70. Genève and Paris. 1821.

HERBST, W. Über Kreuzungen in den Gattung Hypericum mit besonderer Berücksichtigung der Buntblätterigkeit. *Flora* **129**, 235–259. 1935.

KELLER, R. Zur Kenntnis der Sectio Brathys des genus Hypericum. *Bull. Herb. Boiss. sér.* 2 **8**, 175–191. 1908.

STEFANOFF, B. Systematische und geographische Studien über die mediterreen-orientalischen Arten der Gattung Hypericum. *Jahrb. Land. Forstw. Fak. Univ. Sofia* **10**, 19–58. 1932: *loc. cit.* **11**, 139–186. 1933: *loc. cit.* **12**, 69–100. 1934.

CISTACEAE

TUBERARIA

GROSSER, W. *Tuberaria*, in Engler, H. G. A. (Ed.), *Das Pflanzenreich* **14** (IV. 193), 52–61. 1903.

HELIANTHEMUM

GROSSER, W. *Helianthemum*, in Engler, H. G. A. (Ed.), *Das Pflanzenreich* **14** (IV. 193), 61–123. 1903.

PROCTOR, M. C. F. Helianthemum (Biological Flora). *J. Ecol.* **44**, 675–692. 1956.

CISTUS

DANSEREAU, P. M. Monographie du genre Cistus L. *Boissiera* **4**, 1–90. 1939.

GROSSER, W. *Cistus*, in Engler, H. G. A. (Ed.), *Das Pflanzenreich* **14** (IV. 193), 10–32. 1903.

WARBURG, Sir OSCAR and WARBURG, E. F. A preliminary study of the genus Cistus. *J. Roy. Hort. Soc.* **55**, 1–52. 1930.

TAMARICACEAE

TAMARIX

ARENDT, G. *Beiträge zur Kenntnis der Gattung Tamarix*. Pp. 52. Leipzig. 1926.

BATTANDIER, J. A. Revision des Tamarix algériens et description de deux espèces nouvelles. *Bull. Soc. Bot. France* **54**, 232–257. 1907.

BAUM, B. *Monographic revision of the genus Tamarix: final research report*. Pp. iii + 193. Department of Botany, Hebrew University. Jerusalem. 1966.

BAUM, B. and JOVET, P. Étude sur les Tamaris spontanés en France. *Bull. Centre Rech. Sci. Biarritz* **3**, 457–475. 1961.

BUNGE, A. von. *Tentamen Generis Tamaricum Species accuratius definiendi*. Mettieseni. Dorpat. Pp. 52. 1852.

EHRENBURG, C. G. Ueber die Manna-Tamariske nebst algemeinen Bemerkungen über die Tamariscinen. *Linnaea* **2**, 241–344. 1827.

NIEDENZU, F. *De Genere Tamarice*. Pp. 12. Braunsberg. 1895.

RUSANOV, T. Ad studium generis Tamarix L. *Not. Syst. (Leningrad)* **8**, 79–90. 1940.

WILLDENOW, K. L. Beschreibung der Gattung Tamarix. *Abh. Preuss. Akad. Wiss.* **1812–13**, 76–85. 1816.

ELATINACEAE

ELATINE

DUMORTIER, B. C. Examen critique des Elatinées. *Bull. Soc. Bot. Belg.* **11**, 254–274. 1873.

GLUCK, H. *Elatine*, in Pascher, E., *Die Süsswasser-Flora Mitteleuropas* **15**, 299–313. Jena. 1936.

MOESZ, G. Magyarország Elatine-i. *Magyar Bot. Lap.* **7**, 2–35. 1908.

SEUBERT, M. Elatinearum Monographia. *Nova Acta Acad. Leop.-Carol.* **21**, 35–60. 1845.

CARYOPHYLLACEAE

SILENE

CHOWDHURI, P. K. Studies in the genus Silene. *Notes Roy. Bot. Gard. Edinb.* **22**, 221–278. 1957.

HITCHCOCK, C. L. and MAGUIRE, B. A revision of the North American species of Silene. *Univ. Washington Publ. Biol.* **13**, 1–73. 1947.

OTTH, K. A. *Silene*, in de Candolle, A., *Prodromus* **1**, 367–385. 1824.

ROHRBACH, P. Conspectus Systematicus Specierum Generis Silenes. *Ann. Sci. Nat. Sér.* 5 *Bot.* **8**, 369–382. 1867.

ROHRBACH, P. *Monographie der Gattung Silene.* Pp. 247. Leipzig. 1868.

WILLIAMS, F. N. A revision of the genus Silene. *J. Linn. Soc. Bot.* **32**, 1–196. 1896.

LYCHNIS

LAWRENCE, G. H. M. Keys to cultivated plants, 2. The cultivated species of Lychnis. *Baileya* **1**, 105–111, 114. 1953.

DIANTHUS

LAGRAVE, E. T. Essai Monographique sur les Dianthus des Pyrénées Françaises. *Bull. Soc. Agric. Sci. Lit. Pyrénées-Orient* **25**. Pp. 26 + 32 plates. 1881.

LEMPERG, F. Studies on the perennial species of the genus Dianthus, 1. *Meddel. Göt. Bot. Trad.* **11**, 71–134. 1936.

SCHISCHKIN, B. *Dianthus*, in Komarov, V. L., *et al.* (Eds.), *Flora URSS* **6**, 803–861. 1936.

VIERHAPPER, F. Zur Systematik und geographischen Verbreitung einer alpinen Dianthus-Gruppe. *Denkschr. Akad. Wiss. Math.-Nat. Kl. (Wien)* **107**, 1–114. 1898.

WEISSMANN-KOLLMANN, F. A taxonomic study in Dianthus of Palestine, and of the neighbouring countries. *Israel J. Bot.* **14**, 141–148. 1965.

WILLIAMS, F. N. *Enumeration Specierum Varietatumque Generis Dianthus*. Pp. 23. London. 1889.

WILLIAMS, F. N. *Notes on the pinks of western Europe*. Pp. 47. London. 1889.

WILLIAMS, F. N. A monograph of the genus Dianthus. *J. Linn. Soc. Bot.* **29**, 346–478. 1892.

SAPONARIA

SIMMLER, G. Monographie der Gattung Saponaria. *Denkschr. Akad. Wiss. Math.-Nat. Kl.* (*Wien*) **85**, 433–509. 1910.

***PETRORHAGIA* (*KOHLRAUSCHIA*)**

BALL, P. W. and HEYWOOD, V. H. A revision of the genus Petrorhagia. *Bull. Brit. Mus.* (*Bot.*) **3**, 122–172. 1964.

GYPSOPHILA

BARKOUDAH, Y. I. A revision of Gypsophila, Bolanthus, Ankyropetalum and Phryna. *Wentia* **9**, 1–203. 1962.

LAWRENCE, G. H. M. Keys to cultivated plants, 1. The cultivated species of Gypsophila. *Baileya* **1**, 16–18. 1953.

STROH, G. Die Gattung Gypsophila. *Beih. Bot. Centr.* **59**B, 455–477. 1939.

WILLIAMS, F. N. Revision of the specific forms of the genus Gypsophila. *J. Bot.* **27**, 321–329. 1889.

CERASTIUM

BABINGTON, C. C. On the British species of Cerastium. *Mag. Zool. Bot.* **2**, 197–204, 317–319. 1838.

BUSCHMANN, A. Über einige ausdauernde Cerastium-Arten aus der Verwandtschaft des C. tomentosum L. *Fedde. Rep.* **43**, 118–143. 1938.

GARTNER, H. Zur systematischen Anordnung einiger Arten der Gattung Cerastium L. *Fedde. Rep. Beih.* **113**, 1–96. 1939.

GRENIER, J. C. M. *Monographia de Cerastio*. Pp. ii + 95. Vesontione. 1841.

HULTÉN, A. The Cerastium alpinum complex: a case of world-wide introgressive hybridisation. *Svensk Bot. Tidskr.* **50**, 412–495. 1956.

JALAS, J. Notes on Cerastium L. subsect. Perennia Fenzl (Caryophyllaceae). *Arch. Soc. Zool.-Bot. Fenn. Vanamo* **18**, 57–65. 1963.

LONSING, A. Über einjährige europäische Cerastium-Arten aus der Verwandtschaft der Gruppen "Ciliopetala" Fenzl und "Cryptodon" Pax. *Fedde. Rep.* **46**, 139–165. 1939.

MERXMÜLLER, H. Untersuchungen über eine alpine Cerastien-Gruppe. *Ber. Bayer. Bot. Ges.* **28**, 219–238. 1950.

MÖSCHL, W. Über einjährige europäische Arten der Gattung Cerastium (Orthordon-Fugacia-Leiopetala). *Fedde. Rep.* **41**, 153–163. 1936.

MÖSCHL, W. Cerastium semidecandrum Linne, sensu latiore. *Mem. Soc. Brot.* **5**, 5–123. 1949.

Möschl, W. Cerastia Lusitaniae Archipelargorumque Açores et Madeira. *Agron. Lusit.* **13**, 23–66. 1951.

Möschl, W. De Cerastiis Africae septentrionalis. *Mem. Soc. Brot.* **17**, 5–119. 1964.

Murbeck, S. Die nordeuropaeiska formerna af slägtet Cerastium. *Bot. Not.* **1898**, 241–268. 1898.

Sell, P. D. and Whitehead, F. H. Notes on the annual species of Cerastium in Europe. *Fedde. Rep.* **69**, 14–24. 1964.

Söllner, R. Recherches cytotaxonomiques sur le genre Cerastium. *Ber. Schweiz. Bot. Ges.* **64**, 221–354. 1954.

Williams, F. N. Critical notes on some species of Cerastium. *J. Bot.* **36**, 341–344, 382–387. 1898: *loc. cit.* **37**, 116–124, 209–216, 310–315, 474–477. 1899.

Williams, F. N. Énumération provisoire des espèces du genre Cerastium. *Bull. Herb. Boiss.* **6**, 893–904. 1898.

STELLARIA

Murbeck, S. Die nordeuropäischen Formen der Gattung Stellaria. *Bot. Not.* **1899**, 193–218. 1899.

Peterson, D. Stellaria-Studien. Zur Zytologie, Genetik, Ökologie und Systematik der Gattung Stellaria, insbesondere der media-Gruppe. *Bot. Not.* **1936**, 281–419. 1936.

HOLOSTEUM

Gay, J. Holostei, Caryophyllearum Alsinearum Generis, Monographia. *Ann. Sci. Nat. sér.* 2, *Bot.* **4**, 23–44. 1845.

SAGINA

Gibson, S. On British species and varieties of the genus Sagina. *Phyt.* **1**, 177–179. 1842.

Mitzushima, M. A preliminary revision of the genus Sagina of Japan and its adjacent regions. *J. Jap. Bot.* **35**, 77–82, 103–107, 193–200, 257–260. 1960.

Williams, F. N. Revision of the British species of Sagina. *Rep. Bot. Soc. & E.C.* **5**, 190–204. 1918.

MINUARTIA (including ***CHERLERIA***)

Hiern, W. P. Alsine in the British flora. *J. Bot.* **37**, 317–322. 1899.

McNeill, J. Taxonomic studies in the Alsinoideae, 1. Generic and infrageneric groups. *Notes Roy. Bot. Gard. Edinb.* **24**, 79–155. 1962.

Mattfeld, J. Enumeratio specierum generis Minuartia (L.) emend Hiern. *Bot. Jahrb.* **57**, *Beibl.* 126, 27–33. 1921.

MATTFELD, J. Beitrag zur Kenntnis des Systematischen Gliederung und geographischen verbreitung der Gattung Minuartia. *Bot. Jahrb.* **57** *Beibl.* 127,13–63. 1922.

MATTFELD, J. Geographisch genetische Untersuchungen über die Gattung Minuartia (L.) Hiern. *Fedde. Rep. Beih.* **15**, 1–228. 1922.

MOSS, C. E. The genus Alsine. *J. Bot.* **52**, 196–201. 1914.

HONKENYA

MCNEILL, J. Taxonomic studies in the Alsinoideae, 1. Generic and infrageneric groups. *Notes Roy. Bot. Gard. Edinb.* **24**, 79–155. 1962.

POBEDIMOVA, E. G. Revisio generis Honkenya Ehrh. *Not. Syst. (Leningrad)* **20**, 142–162. 1960.

MOEHRINGIA

BORNMÜLLER, J. Zur Gattung Moehringia. *Fedde. Rep.* **16**, 183–186. 1919.

ARENARIA

MCNEILL, J. Taxonomic studies in the Alsinoideae, 1. Generic and infrageneric groups. *Notes Roy. Bot. Gard. Edinb.* **24**, 79–155. 1962.

OSTENFELD, C. H. and DAHL, O. De nordiske former av Kollektivarten Arenaria ciliata L. *Nytt Mag. Nat.* **55**, 215–225. 1917.

WILLIAMS, F. N. On the genus Arenaria Linn. *Bull. Herb. Boiss.* **3**, 593–603. 1895.

WILLIAMS, F. N. A revision of the genus Arenaria Linn. *J. Linn. Soc. Bot.* **33**, 326–437. 1898.

SPERGULARIA

KINDBERG, N. C. *Synoptisk framställning af växtslägtet Lepigonum.* Pp. 16. Upsala. 1856.

KINDBERG, N. C. Monographia generis Lepigonorum. *Nova Acta Regiae Soc. Sci. Upsal. sér.* 3 **4** (7). Pp. 44. 1863.

LEBEL, E. Revision du genre Spergularia. *Mém. Soc. Imp. Sci. Nat. Cherbourg* **14**, 43–75. 1868.

MONNIER, P. Introduction à une revision du genre Spergularia (Pers.) Presl au Maroc. *Bull. Soc. Sci. Nat. Maroc* **35**, 145–163. 1955.

ROSSBACH, R. P. Spergularia in North and South America. *Rhodora* **42**, 57–83, 105–143, 158– 93, 203–213. 1940.

ILLECEBRACEAE

HERNIARIA

HERMANN, F. Übersicht über die Herniaria-Arten des Berliner Herbars. *Fedde. Rep.* **42**, 203–224. 1937.

WILLIAMS, F. N. A systematic revision of the genus Herniaria. *Bull. Herb. Boiss.* **4**, 556–570. 1896.

PARONYCHIA

Borhidi, A. The variability range of Paronychia cephalotes (M.B.) Bess. and new data to the knowledge of south European Paronychia species. *Acta Bot. Hung.* **12**, 33–40. 1966.

Core, L. The North American species of Paronychia. *Amer. Midl. Nat.* **26**, 369–395. 1941.

SCLERANTHUS

Rössler, W. Die Scleranthus-Arten Reichenbach's. *Ann. Nat. Mus.* (*Wien*) **57**, 97–129. 1950.

Rössler, W. Scleranthi Lusitaniae. *Agron. Lusit.* **15**, 97–138. 1953.

Rössler, W. Neues aus dem Scleranthus-Nachlass Reichenbach's. *Phyton* **5**, 222–227. 1954.

Rössler, W. Die Scleranthus-Arten Österreichs und seiner Nachbarländer. *Österr. Bot. Zeitschr.* **102**, 30–72. 1955.

Sell, P. D. Notes on the European species of Scleranthus. *Fedde. Rep.* **68**, 167–169. 1963.

Vierhapper, F. Die systematische Stellung der Gattung Scleranthus. *Österr. Bot. Zeitschr.* **57**, 41–47, 91–96. 1907.

PORTULACACEAE

MONTIA

Clason, E. W. Montia fontana in Nederland. *Acta Bot. Neerl.* **4**, 242–272. 1955.

Lawalrée, A. Revision des Montia de Belgique de l'herbier du Jardin Botanique de l'État. *Bull. Jard. Bot. Bruxelles* **30**, 97–104. 1960.

Walters, S. M. Montia fontana L. *Watsonia* **3**, 1–6. 1953.

PORTULACA

Poellnitz, K. von. Versuch einer Monographie der Gattung Portulaca. *Fedde. Rep.* **37**, 240–320. 1934.

AMARANTHACEAE

AMARANTHUS

Aellen, P. *Die Amaranthaceen Mittel-europas*, in Hegi, G., *Illustr. Fl. Mitteleur.* Edition 2 **3** (2), 465–516. 1959.

Brenan, J. P. M. Amaranthus in Britain. *Watsonia* **4**, 261–280. 1961.

d'Alleizette, C. and Aellen, P. Bemerkenswerte Amaranthus-Funde aus Frankreich. *Candollea* **14**, 159–162. 1953.

Kloos, A. W. *Amaranthus*, in Robyns, A., *Fl. Gén. Belg.* (*Spermat.*) **1**, 306–335. 1953.

Priszter, S. Revisio critica specierum generis Amaranthi L. in Hungaria. *Agrártud. Egyetem Kert Kar. Közl.* **2**, 121–262. 1953.

PRISZTER, S. Über die bisher bekannten Bastarde der Gattung Amaranthus. *Bauhinia* **1**, 126–135. 1958.

SAUER, J. Revision of the dioecious Amaranths. *Madroño* **13**, 5–46. 1955.

THELLUNG, A. *Amaranthus*, in Ascherson, P. and Graebner, P., *Synopsis der Mitteleur. Fl.* **5**(1), 225–356. 1914.

ULINE, E. B. and BRAY, W. L. A preliminary synopsis of the North American species of Amaranthus. *Bot. Gaz.* **19**, 267–272, 313–320. 1894.

CHENOPODIACEAE

MOQUIN, A. Mémoires sur la famille des Chénopodées. *Ann. Sci. Nat.* **23**, 274–325. 1831.

MOQUIN-TANDON, A. Conspectus Generum Chenopodearum (Atriplicearum Juss. et Chenopodearum DC. Gen.). *Ann. Sci. Nat. sér.* 2 **4**, 209–218. 1835.

MOQUIN-TANDON, A. *Chenopodearum Monographica Enumeratio.* Pp. xi + 168. Paris. 1840.

CHENOPODIUM

AELLEN, P. Beiträge zur Systematik der Chenopodium-Arten Amerikas, vorwiegend auf Grund der Sammlung des United States National Museum in Washington, D.C. *Fedde. Rep.* **26**, 31–64, 119–160, 1929.

AELLEN, P. Die Wooladventiven Chenopodien Europas. *Verh. Naturf. Ges. Basel* **41**, 77–104. 1930.

AELLEN, P. *Chenopodium*, in Hegi, G., *Illustr. Fl. Mitteleur.* Edition 2. **3** (2), 569–659. 1960.

AELLEN, P. and JUST, T. Key and synopsis of the North American species of the genus Chenopodium. *Amer. Midl. Nat.* **30**, 47–76. 1943.

KOWAL, T. Kluczdo oznaczania nasion rodzajów Chenopodium L. i Atriplex L. *Monogr. Bot.* **1**, 87–163. 1953.

MOQUIN-TANDON, A. Conspectus Generum Chenopodearum (Atriplicearum Juss. et Chenopodearum DC. gen.). *Ann. Sci. Nat. sér.* 2 **4**, 209–218. 1835.

MOQUIN-TANDON, A. *Chenopodearum Monographica Enumeratio.* Pp. xi + 168. Paris. 1840.

AXYRIS

MOQUIN-TANDON, A. *Chenopodearum Monographica Enumeratio.* Pp. xi + 168. Paris. 1840.

BETA

AELLEN, P. Die Orientalischen Beta-Arten. *Ber. Schweiz. Bot. Ges.* **48**, 470–484. 1938.

HELM, J. Über die historische Entwicklung der Gliederung von Beta vulgaris L. in Untersippen und deren Nomenklatur. *Kulturpflanze* **5**, 55–74. 1957.

MOQUIN-TANDON, A. *Chenopodearum Monographica Enumeratio*. Pp. xi + 168. Paris. 1840.

TJEBBER, K. The wild beets of the North Sea region. *Bot. Not.* **1933**, 305–315. 1933.

SPINACIA

AELLEN, P. Beitrag zur Kenntnis von Spinacia L. *Ber. Schweiz. Bot. Ges.* **48**, 485–490. 1938.

ATRIPLEX

BABINGTON, C. C. Monograph of the British Atripliceae. *Trans. Proc. Bot. Soc. Edinb.* **1**, 1–17. 1841.

BROWN, G. D. Taxonomy of American Atriplex. *Amer. Midl. Nat.* **55**, 199–210. 1956.

KOWAL, T. Kluczdo oznaczania nasion rodzajów Chenopodium L. i Atriplex L. *Monogr. Bot.* **1**, 87–163. 1953.

MOQUIN-TANDON, A. Conspectus Generum Chenopodearum (Atriplicearum Juss. et Chenopodearum DC. Gen.). *Ann. Sci. Nat. sér.* 2 **4**, 209–218. 1835.

MOQUIN-TANDON, A. *Chenopodearum Monographica Enumeratio*. Pp. xi + 168. Paris. 1840.

HALIMIONE

MOQUIN-TANDON, A. Conspectus Generum Chenopodearum (Atriplicearum Juss. et Chenopodearum DC. Gen.). *Ann. Sci. Nat. sér.* 2 **4**, 209–218. 1835.

MOQUIN-TANDON, A. *Chenopodearum Monographica Enumeratio*. Pp. xi + 168. Paris. 1840.

SUAEDA

MOQUIN, A. Mémoires sur la famille des Chénopodés. *Ann. Sci. Nat.* **23**, 274–325. 1831.

MOQUIN-TANDON, A. *Chenopodearum Monographica Enumeratio*. Pp. xi + 168. Paris. 1840.

SALSOLA

MOQUIN-TANDON, A. *Chenopodearum Monographica Enumeratio*. Pp. xi + 168. Paris. 1840.

KOCHIA

AELLEN, P. Kochia Roth. *Mitt. Basle. Bot. Ges.* **2**, 4–16. 1954.

SALICORNIA

BALL, P. W. A taxonomic review of Salicornia in Europe. *Fedde. Rep.* **69** 1–8. 1964.

BALL, P. W. and TUTIN, T. G. Notes on annual species of Salicornia in Britain. *Watsonia* **4**, 193–205. 1959.
KÖNIG, D. Beiträge zur Kenntnis der deutschen Salicornien. *Mitt. Fl.-Soz. Arbeitsgem. nov. ser.* **8**, 5–58. 1960.
MOSS, C. E. Some species of Salicornia. *J. Bot.* **49**, 177–185, 233–234. 1911.
NANNFELDT, J. A. Något om släktet Salicornia i Sverige. *Svensk Bot. Tidskr.* **49**, 97–109. 1955.
SOÓ, R. von. Über südosteuropäische Salicornien. *Acta Bot. Hung.* **6**, 397–403. 1960.
UNGERN-STERNBERG, FRANZ, Baron. *Versuch einer Systematik der Salicornieen.* Pp. xiv + 114. Dorpat. 1866.

PHYTOLACCACEAE

PHYTOLACCA

WALTER, H. *Phytolacca*, in Engler, H. G. A. (Ed.), *Das Pflanzenreich* **39** (IV.83), 36–63. 1909.

TILIACEAE

TILIA

BURRET, M. Beiträge zur Kenntnis der Tiliaceen. *Notizbl. Bot. Gart. Berlin* **9**, 592–880. 1926.
ENGLER, V. *Monographie der Gattung Tilia.* Pp. 159. Breslau. 1909.
SPACH, E. Revisio Generis Tiliarum. *Ann. Sci. Nat. sér.* 2 **2**, 331–347. 1834.
WAGNER, J. Lindenstudien. *Magyar Bot. Lap.* **31**, 55–60. 1932.

MALVACEAE

BAKER, E. G. Synopsis of genera and species of Malveae. *J. Bot.* **28**, 15–18, 140–145, 207–213, 239–243, 339–343, 367–371. 1890: *loc. cit.* **29**, 49–53, 164–172, 362–366. 1891: *loc. cit.* **30**, 71–78, 136–142, 235–240, 290–296, 324–332. 1892: *loc. cit.* **31**, 68–76, 212–217, 267–273, 334–338, 361–368. 1893.
BAKER, E. G. Supplement to synopsis of Malveae. *J. Bot.* **32**, 35–38. 1894.

MALVA

BAKER, E. G. Synopsis of genera and species of Malveae. *J. Bot.* **28**, 242–243, 339–343, 367–371. 1890.
*SLAVIK, B. *Připavná studie k monografii československých slézu (Diplomá prace).* Pp. 228 + 36 Tab. Praha. 1958.

LAVATERA

BAKER, E. G. Synopsis of genera and species of Malveae. *J. Bot.* **28**, 210–213, 239–241. 1890.

ALTHAEA

BAKER, E. G. Synopsis of genera and species of Malveae. *J. Bot.* **28**, 140–145, 207–209. 1890.

ABUTILON

BAKER, E. G. Synopsis of genera and species of Malveae. *J. Bot.* **31**, 71–76, 212–217, 267–273, 334–338, 367. 1893.

KEARNEY, T. H. A tentative key to the North American species of Abutilon. *Leafl. West. Bot.* **7**, 241–254. 1955.

HIBISCUS

HOCHREUTINER, B. P. G. Revision du genre Hibiscus. *Ann. Conserv. Gard. Bot. Genève* **4**, 23–191. 1900.

SIDA

BAKER, E. G. Synopsis of genera and species of Malveae. *J. Bot.* **30**, 138–142, 235–240, 290–296, 324–332. 1892.

GANDOGER, M. Le genre Sida (Malvacées). *Bull. Soc. Bot. France* **71**, 627–633. 1924.

KEARNEY, T. H. A tentative key to the North American species of Sida. *Leafl. West. Bot.* **6**, 138–150. 1954.

SIDALCEA

BAKER, E. G. Synopsis of genera and species of Malveae. *J. Bot.* **29**, 51–53. 1891.

ROUSH, E. M. F. A monograph of the genus Sidalcea. *Ann. Missouri Bot. Gard.* **18**, 117–244. 1931.

WISSADULA

BAKER, E. G. Synopsis of genera and species of Malveae. *J. Bot.* **31**, 69–71. 1893.

TROPAEOLACEAE

TROPAEOLUM

BUCHENAU, F. Beiträge zur Kenntnis der Gattung Tropaeolum. *Bot. Jahrb.* **15**, 180–259. 1892.

LINACEAE

LINUM

CIFERRI, R. *La sistematica de Lino. Secunde Wulff and Elladi.* Pp. 203. Bologna. 1949.

NESTLER, H. Beiträge zur systematischen Kenntnis der Gattung Linum. *Beih. Bot. Centr.* **50** (2), 497–551. 1933.

GERANIACEAE

Sweet, R. *Geraniaceae*. 4 vols and supplement. London. 1820–30.

GERANIUM

Carolin, R. C. The genus Geranium in the south-western Pacific area. *Proc. Linn. Soc. New South Wales* **89**, 326–361. 1965.

Jones, G. N. and Jones, F. F. A revision of the perennial species of Geranium in the United States and Canada. *Rhodora* **45**, 5–26, 32–53. 1943.

Knuth, R. *Geranium*, in Engler, H. G. A. (Ed.), *Das Pflanzenreich* **53** (IV. 129), 43–221, 575–583. 1912.

ERODIUM

Andreas, C. M. De inheemsche Erodia van Nederland. *Nederl. Kruidk. Arch.* **54**, 138–231. 1947.

Baker, E. G. and Salmon, C. E. Some segregates of Erodium cicutarium L'Hérit. *J. Bot.* **58**, 123–127. 1921.

Brumhard, P. Monographische Übersicht der Gattung Erodium. *Arbeit. Bot. Gart. Univ. Breslau* **1905**. Pp. 59. 1905.

Carolin, R. C. The species of the genus Erodium L'Hér. endemic to Australia. *Proc. Linn. Soc. New South Wales* **83**, 92–100. 1958.

Knuth, R. *Erodium*, in Engler, H. G. A. (Ed.), *Das Pflanzenreich* **53** (IV. 129). 221–290, 583–587. 1912.

Larsen, K. Cytological and experimental studies on the genus Erodium with special references to the collective species E. cicutarium (L.) L'Hér. *Biol. Medd.* **23** (6), 1–25. 1958.

OXALIDACEAE

OXALIS

Eiten, G. Taxonomy and regional variation of Oxalis section Corniculatae, 1. Introduction, keys and synopsis of the species. *Amer. Midl. Nat.* **69**, 257–309. 1963.

Ingram, J. The cultivated species of Oxalis, 1. The caulescent species. *Baileya* **6**, 23–32. 1958: 2. The acaulescent species, *loc. cit.* **7**, 11–23. 1959.

Jacquin, N. J. von. *Oxalis: Monographia Iconibus Illustrata*. Pp. 119. Viennae. 1794.

Knuth, R. *Oxalis*, in Engler, H. G. A. (Ed.), *Das Pflanzenreich* **95** (IV. 130), 43–389. 1930.

Young, D. P. Oxalis in the British Isles. *Watsonia* **4**, 51–69. 1958.

Zuccarini, J. G. Monographie der Amerikanischen Oxalis-Arten. *Denkschr. Akad. München* **9**, 125–184. 1825.

Zuccarini, J. G. Nachtrag zu der Monographie der amerikanischen Oxalis-Arten. *Denkschr. Akad. München Abhand.* **1**, 177–276. 1831.

ACERACEAE

ACER

PAX, F. Monographie der Gattung Acer. *Bot. Jahrb.* **6**, 287–374. 1885: *loc. cit.* **7**, 177–263. 1885–86: *loc. cit.* **11**, 72–83. 1889. Supplement. *Loc.cit.* **16**, 392–404. 1893.

PAX, F. *Acer*, in Engler, H. G. A. (Ed.), *Das Pflanzenreich* **8** (IV. 163), 6–80. 1902.

SPACH, E. Revisio Generis Acerum. *Ann. Sci. Nat. sér.* 2 **2**, 160–180. 1834.

AQUIFOLIACEAE

LOESENER, T. Monographia Aquifoliacearum. *Abh. K. Leop.-Carol. Deutsch. Akad. Nat.* **78**, 1–598. 1900: *loc. cit.* **89**, 1–313. 1908.

LOESENER, T. Über die Aquifoliaceen besonders über Ilex. *Mitt. Deutsch. Dendr. Ges.* **1919**, 1–66. 1919.

ILEX

DENGLER, H. W. (Ed.). *Handbook of Hollies.* Pp. vi + 193. London. 1957.

LOESENER, T. Monographia Aquifoliacearum. *Abh. K. Leop.-Carol. Deutsch. Akad. Nat.* **78**, 1–598. 1900: *loc. cit.* **89**, 1–313. 1908.

LOESENER, T. Über die Aquifoliaceen besonders über Ilex. *Mitt. Deutsch. Dendr. Ges.* **1919**, 1–66. 1919.

CELASTRACEAE

EUONYMUS

BARÁTH, Z. Hazai Euonymus-ainkról. *Bot. Közl.* **46**, 235–250. 1956.

BLAKELOCK, R. A. A synopsis of the genus Euonymus. *Kew Bull.* **1951**, 210–292. 1952.

KLOKOV, M. De Euonymo Europaea auct Florae URSS. *Not. Syst. (Leningrad)* **19**, 274–329. 1959.

BUXACEAE

BAILLON, H. E. *Monographie des Buxacées et de Stylocerées.* Pp. 87. Paris. 1859.

BUXUS

BAILLON, H. E. *Monographie des Buxacées et de Stylocerées.* Pp. 87. Paris. 1859.

RHAMNACEAE

RHAMNUS

GRUBOV, V. I. Monographic sketch of the genus Rhamnus L. S.l. *Acta Inst. Bot. Acad. Sci. URSS ser.* 1 **8**, 243–423. 1949.

VITACEAE

PLANCHON, J. E. *Monographie des Ampélidées vraies*, in de Candolle A., *Monogr. Phanerog.* **V** (2), 306–654. 1887.

VITIS

PLANCHON, J. E. *Monographie des Ampélidées vraies*, in de Candolle, A., *Monogr. Phanerog.* **V** (2), 306–654. 1887.

PARTHENOCISSUS

PLANCHON, J. E. *Monographie des Ampélidées vraies*, in de Candolle, A., *Monogr. Phanerog.* **V** (2), 306–654. 1887.

REHDER, A. Die amerikanischen Arten der Gattung Parthenocissus. *Mitt. Deutsch. Dendr. Ges.* **1905**, 129–136. 1905.

LEGUMINOSAE

LUPINUS

AGARDH, J. G. *Synopsis Generis Lupini.* Pp. xiv + 43 + 2 plates. Lundae. 1835.

PHILLIPS, L. L. A revision of the perennial species of Lupinus in North America. *Res. Stud. State Coll. Washington* **23**, 161–204. 1955.

SMITH, C. P. *Species Lupinorum.* Pp. 768. Saratoga. 1938–53.

GENISTA

GIBBS, P. E. A revision of the genus Genista L. *Notes Roy. Bot. Gard. Edinb.* **27**, 11–99. 1966.

SPACH, E. Revisio Generis Genista. *Ann. Sci. Nat. sér.* 3 *Bot.* **2**, 237–279. 1844: *loc. cit.* **3**, 102–158. 1845.

SPARTIUM

SPACH, E. Monographia Generis Spartium. *Ann. Sci. Nat. sér.* 2 **19**, 285–297. 1843.

ULEX

RIKLI, M. Der Mitteleuropäischen Arten der Gattung Ulex. *Ber. Schweiz. Bot. Ges.* **8**, 1–15. 1898.

ROTHMALER, W. Revision der Genisteen, 1. Monographie der Gattung um Ulex. *Bot. Jahrb.* **72**, 69–116. 1941.

VICIOSO, C. Revisión del género Ulex en España. *Bol. Inst. Forest. Invest. Exp.* **33** (80). Pp. 60. 1962.

SAROTHAMNUS

ROTHMALER, W. Die Gliederung der Gattung Cytisus. *Fedde. Rep.* **53**, 137–150. 1944.

CYTISUS

HOLUBOVÁ-KLÁSKOVÁ, A. Bemerkungen zur Gliederung der Gattung Cytisus L. s.l. *Acta Univ. Carolinae* 1964, *suppl.* **2**, 1–24. 1964.

PSORALEA

OCKENDON, D. J. A taxonomic study of Psoralea subgenus Pediomelum (Leguminosae). *Southwest. Nat.* (*Texas*) **10**, 81–124. 1965.

ONONIS

ERDTMAN, J. Zur Verbreitung und Taxonomie der Gattung Ononis in Nord-ost-Deutschland. *Fedde. Rep.* **69**, 103–131. 1964.

ŠIRJAEV, G. Generis Ononis L. revisio critica. *Beih. Bot. Centr.* **49** (2), 381–665. 1932.

MEDICAGO

CASELLAS, J. El género Medicago L. en España. *Collect. Bot.* **6**, 183–291. 1962.

HEYN, C. C. The annual species of Medicago. *Scripta Hier.* **12**, 1–154. 1963.

KOŽUHAROV, S. Vidovete na rod Medicago L. (Ljucerna) v Bălgarija. *Izvest. Bot. Inst.* (*Sofia*) **15**, 119–188. 1965.

NÈGRE, R. Les Luzernes du Maroc. *Trav. Inst. Sci. Chérif. sér. Bot.* **5**. Pp. xxi + 119. 1956.

NÈGRE, R. Révision des Medicago d'Afrique du Nord. *Bull. Soc. Hist. Afr. Nord.* **50**, 267–314. 1959.

QUINLIVAN, B. J. The naturalised and cultivated annual Medics of Western Australia. *J. Agric. W. Australia* **6**, 532–543. 1965.

URBAN, I. Prodromus einer Monographie der Gattung Medicago L. *Verh. Bot. Ver. Brandenb.* **15**, 1–85. 1873.

URBAN, I. Die Medicago-Arten Linne's. *Ber. Deutsch. Bot. Ges.* **1**, 256–262. 1883.

VAN OOSTSTROOM, S. J. and REICHGELT, T. J. Het geslacht Medicago in Nederland en België. *Acta Bot. Neerl.* **7**, 90–123. 1958.

VASSILECZENKO, I. T. Monographic sketch of the perennial species of the genus Medicago L. sect. Falcago Rchb. *Acta Inst. Bot. Acad. Sci. U.R.S.S. ser.* 1 **8**, 1–420. 1949.

MELILOTUS

ISELY, D. Keys to the sweet clovers (Melilotus). *Proc. Iowa Acad. Sci.* **61**, 119–131. 1954.

KITA, F. Studies on the genus Melilotus (sweet clover) with special reference to interrelationship amongst species from a cytological point of view. *J. Fac. Agric. Hokkaido Univ.* **54** (2), 23–122. 1965.

SCHULZ, O. E. Monographie der Gattung Melilotus. *Bot. Jahrb.* **29**, 660–737. 1901.

TRIFOLIUM

BOBROV, E. G. Vidy kleverov SSSR. *Acta Inst. Bot. Acad. Sci. URSS ser* 1 **6**, 164–336. 1947.

*GIBELLI, G. and BELLI, S. Rivista critica delle specie di Trifolium italiane comparate con quelle del resto d'Europa e delle regioni circummediter rannee. *Mem. Accad. Sci. Torino.* 1890–92.

HERMANN, F. Übersicht über die europäischen Rotten und Arten und einige andere Arten der Gattung Trifolium. *Fedde. Rep.* **39**, 332–351. 1936.

HOSSAIN, M. A revision of Trifolium in the nearer East. *Notes Roy. Bot. Gard. Edinb.* **23**, 387–481. 1961.

KATZNELSON, J. and MORLEY, F. H. W. A taxonomic revision of Sect. Calycomorphum of the genus Trifolium, 1. The geocarpic species. *Israel J. Bot.* **14**, 112–134. 1965.

LOJACONO, M. Clavis specierum Trifoliorum. *Nuov. Giorn. Bot. Ital.* **15**, 225–278. 1883.

McDERMOTT, L. F. *An Illustrated Key to the North American species of Trifolium.* Pp. 325. San Francisco. 1910.

SAVI, G. *Observationes in varias Trifoliorum species.* Pp. 116. Florentiae. 1810.

VICIOSO, C. Tréboles españoles. Revisión del género Trifolium. *Anal. Inst. Bot. Cav.* **10**, 347–398. 1952: *loc. cit.* **11** (1), 289–383. 1953.

TRIGONELLA

GASPARRINI, G. *Revisio generis Trigonellae et super nonnullis aliis plantis annotationes.* Pp. 9. Neapoli. 1852.

ŠIRJAEV, G. Generis Trigonella L. revisio critica. *Publ. Fac. Sci. Univ. Masaryk (Brno)* **1**. Pp. 57. 1928: *loc. cit.* **2**. Pp. 37. 1929: *loc. cit.* **3**. Pp. 31. 1930: *loc. cit.* **4**. Pp. 33. 1931: *loc. cit.* **5**. Pp. 48. 1932: *loc. cit.* **6**. Pp. 37. 1933.

ŠIRJAEV, G. Die Entwicklungsgeschichte der Gattung Trigonella. *Bull. Ass. Russe Rech. Sci. (Prague)* **2**, 135–162. 1935.

VASSILCZENKO, I. T. Obzor vidov roda Trigonella L. *Acta Inst. Bot. Acad. Sci. URSS ser.* 1 **10**, 124–269. 1953.

ANTHYLLIS

BECKER, W. Bearbeitung der Anthyllis Sektion-Vulneraria DC. *Beih. Bot. Centr.* **27** (2), 256–287. 1910.

BECKER, W. Anthyllisstudien. *Beih. Bot. Centr.* **29** (2), 16–40. 1912.

JALAS, J. Rassentaxonomische Probleme im Bereich des Anthyllis vulneraria L.-Komplexes in Belgien. *Bull. Jard. Bot. Bruxelles* **27**, 405–416. 1957.

MARSDEN-JONES, E. M. and TURRILL, W. B. Studies in the variation of Anthyllis vulneraria. *J. Gen.* **27**, 261–285. 1933.

MARSDEN-JONES, E. M. and TURRILL, W. B. Notes on the taxonomy of British material of Anthyllis vulneraria. *J. Bot.* **71**, 207–213. 1933.

ROTHMALER, W. Westmediterrane Arten der Sektion Vulneraria DC. der Gattung Anthyllis L. *Fedde. Rep.* **50**, 177–192, 233–245. 1941.

SAGORSKI, E. Über den Formenkreis der Anthyllis vulneraria L. *Allgem. Bot. Zeitschr.* **14**, 40–43, 55–58, 89–93, 124–134, 154–157, 172–175, 184–189, 204–205. 1908: *loc. cit.* **15**, 7–11, 19–23. 1909.

DORYCNIUM

RIKLI, M. Die schweizerischen Dorycnien. *Ber. Schweiz. Bot. Ges.* **10**, 10–44. 1900.

RIKLI, M. Die Gattung Dorycnium Vill. *Bot. Jahrb.* **31**, 314–404. 1901.

LOTUS

BRAND, A. Monographie der Gattung Lotus. *Bot. Jahrb.* **25**, 166–232. 1898.

LARSEN, K. and ŽERTOVÁ, A. On the variation pattern of Lotus corniculatus in eastern Europe. *Bot. Tidsskr.* **59**, 177–194. 1963.

LARSEN, K. and ŽERTOVÁ, A. The Australian Lotus species. *Fedde. Rep.* **72**, 1–18. 1965.

MINIAEV, H. A. De speciebus generis Lotus L. in regionibus occidentali-septentrionalibus partis Europeae URSS crescentibus, 1. *Not. Syst. (Leningrad)* **18**, 119–141. 1957.

UJHELYI, J. Études taxonomiques sur le groupe Lotus corniculatus L. sensu lato. *Ann. Mus. Hung.* **52**, 185–200. 1960.

ŽERTOVÁ, A. A. Studie über die tschechoslowakischen Arten der Gattung Lotus L. *Preslia* **33**, 17–35. 1961: *Sborn. národ. Mus. Praha* **18**, 107–119. 1962: *Acta Hort. Bot. Pragensis* **1963**, 80–83. 1964: *Preslia* **38**, 36–47. 1966.

TETRAGONOLOBUS

DAVEU, M. J. Note sur quelques Lotus de la section Tetragonolobus. *Bull. Soc. Bot. France* **43**, 358–369. 1896.

COLUTEA

BROWICZ, K. The genus Colutea L. A monograph. *Monogr. Bot.* **14**, 3–135. 1963.

ASTRAGALUS

BUNGE, A. Generis Astragali species gerontogeae. *Mém. Acad. Sci. St. Pétersb. (Sci. Phys. Math.) sér.* 7 **11** (6). Pp. 140. 1868: *loc. cit.* **15** (1). Pp. 254. 1869.

OXYTROPIS

BUNGE, A. Species generis Oxytropis DC. *Mém. Acad. Sci. St. Pétersb. (Sci. Phys. Math.) sér.* 7 **22** (1). Pp. 166. 1874.

GUTERMANN, W. and MERXMÜLLER, H. Die europäischen Sippen von Oxytropis sectio Oxytropis. *Mitt. Bot. Staats. München* **4**, 199–275. 1961.

LEINS, P. and MERXMÜLLER, H. Zur Gliederung der Oxytropis campestris-Gruppe. *Mitt. Bot. Staats. München* **6**, 19–31. 1966.

CORONILLA

UHROVÁ, A. Revision der Gattung Coronilla. *Beih. Bot. Centr.* **53**B, 1–174. 1935.

HIPPOCREPIS

Hrabětová-Uhrová, A. Generis Hippocrepis revisio. *Acta Acad. Sci. Nat. Morav.-Siles.* **21** (4), 1–54. 1949: *loc. cit.* **22**, 99–158, 219–250, 331–356. 1950.

ONOBRYCHIS

Širjaev, G. Onobrychis Generis revisio critica. *Publ. Fac. Sci. Univ. Masaryk (Brno)* **56**, 1–195. 1925: *loc. cit.* **76**, 1–165. 1926.

Širjaev, G. Supplementum ad monographium "Onobrychis generis revisio critica". *Bull. Soc. Bot. Bulg.* **4**, 7–24. 1931.

VICIA

Guinea, E. *Estudio Botánico de las Vezas y Arvejas Españolas.* Pp. 223. Madrid. 1953.

Hermann, F. J. Vetches of the United States—native, naturalised and cultivated. *U.S. Dept. Agric. Handb.* **168**. Pp. 84. 1960.

Kostrakiewicz, K. Studia systematyczne nad polskimi gatunkami rodzaju Vicia L. *Bull. Acad. Polon. Sci. Lett.* **27**, 1–71. 1951.

Mettin, D. and Hanelt, P. Cytosystematische Untersuchungen in der Artengruppe um Vicia sativa L. *Kulturpflanze* **12**, 163–225. 1964.

LATHYRUS

Bässler, M. Die Stellung des Subgenus Orobus (L.) Baker in der Gattung Lathyrus L. und seine systematische Gliederung. *Fedde. Rep.* **72**, 69–97. 1966.

Czefranova, Z. Conspectus criticus Specierum Sectionis Orobus (L.) Gr. et Godr. Generis Lathyrus L. Florae URSS. *Nov. Syst. Plant. Vasc.* **1965**, 152–167. 1965.

Ginzberger, A. Über einige Lathyrus-Arten aus der Sektion Eulathyrus und ihre geographische Verbreitung. *Sitz.-Ber. Akad. Wiss. Wien, Abt.* I **105**, 281–352. 1896.

Senn, H. A. Experimental data for a revision of the genus Lathyrus. *Amer. J. Bot.* **25**, 67–78. 1938.

White, T. G. A preliminary revision of the genus Lathyrus in North and Central America. *Bull. Torrey Bot. Club* **21**, 444–458. 1894.

SCORPIURUS

Heyn, C. H. and Raviv, V. Experimental taxonomic studies in the genus Scorpiurus (Papilionaceae). *Bull. Torrey Bot. Club* **93**, 259–267. 1966.

CICER

aubert, M. and Spach, E. Monographia generis Cicer. *Ann. Sci. Nat. sér.* 2 **18**, 223–235. 1842.

GLYCINE

HERMANN, F. J. A revision of the genus Glycine and its immediate allies. *U.S. Dept. Agric. Techn. Bull.* **1268**, Pp. 82. 1962.

PHASEOLUS

HASSLER, E. Revisio Specierum austro-Americaniae generis Phaseoli L. *Candollea* **1**, 417–472. 1923.

ROSACEAE

TRATTINNICK, L. *Rosacearum Monographia.* 4 vols. Vindobonae. 1823–24.

SPIRAEA

CAMBESSÈDES, J. *Monographie du genre Spiraea.* Pp. 58. Paris. 1824.

FILIPENDULA

SHIMIZU, T. Taxonomical notes on the genus Filipendula (Rosaceae). *J. Fac. Text. Techn. Shinshu Univ.* **26**A (10), 1–31. 1961.

HOLODISCUS

LEY, A. A taxonomic revision of the genus Holodiscus (Rosaceae). *Bull. Torrey Bot. Club* **70**, 275–288. 1943.

RUBUS

BABINGTON, C. C. *The British Rubi.* Pp. viii + 305. London. 1869.

BEIJERINCK, W. The Rubus-flora of Belgium and the Netherlands, its study and its problems. *Biol. Jaarb.* **19**, 28–51. 1952.

BEIJERINCK, W. On the habit, ecology and taxonomy of the brambles of the Netherlands. *Acta Bot. Neerl.* **1**, 523–546. 1953.

BEIJERINCK, W. Rubi Neerlandica. *Verh. Kon. Akad. Wetenschapp.* **51**, 1–156. 1956.

FOCKE, W. O. *Synopsis Ruborum Germaniae.* Pp. 434. Bremen. 1877.

FOCKE, W. O. Species Ruborum. Monographiae generis rubi prodromus. *Bibl. Bot.* **72** (1 & 2). Pp. 223. 1910–11: *loc. cit.* **83**. Pp. 274. 1914.

GENEVIER, L. G. Essai Monographique sur les Rubus du Bassin de la Loire. *Mém. Soc. Acad. Maine Loire* **24**. Pp. 346. 1868: *Premier Supplément. Loc. cit.* **28**. Pp. 96. 1872. Edition 2. Pp. 394. Paris. 1880.

GUSTAFSSON, Å. The genesis of the European blackberry flora. *Lunds Univ. Årsskr. ser.* 2 **39** (6), 1–199. 1943.

HRUBY, J. Beiträge zur Systematik der Gattung Rubus L. *Fedde. Rep.* **33**, 379–392. 1934: *loc. cit.* **36**, 352–384. 1935: *loc. cit.* **38**, 172–180. 1935.

LEGRAIN, J. Catalogue des Ronces de Belgique. *Bull. Soc. Bot. Belg.* **89**, 21–34. 1957.

Müller, P. J. and Lefèvre, V. Versuch einer monographischen Darstellung der gallo-germanischen Arten der Gattung Rubus. *Pollichia* **17**. 1859.

Rogers, W. M. *Handbook of British Rubi*. Pp. ix + 111. London. 1900.

Sudre, H. *Rubi Europae vel Monographia iconibus illustratia Ruborum Europae*. Pp. 305 + 215 plates. Paris. 1908–13.

Watson, W. C. R. *Handbook of the Rubi of Great Britain and Ireland*. Pp. xi + 274. Cambridge. 1958.

Weihe, K. E. A. and Nees von Esenbeck, C. G. *Rubi Germanici, descripti et figuris illustrati*. Pp. 120 + 49 plates. Elberfeld. 1822–27.

POTENTILLA

Ascherson, P. and Graebner, P. *Potentilla*, in *Syn. Mitteleur. Fl.* **6** (1), 864–872. 1901.

Borhidi, A. and Isépy, I. Taxa et combinationes novae generis Potentilla L. *Acta Bot. Hung.* **11**, 298–302. 1965.

Bowden, W. M. Cytotaxonomy of Potentilla fruticosa, allied species and cultivars. *J. Arnold Arb.* **38**, 381–388. 1957.

Lehmann, J. G. C. *Monographia generis Potentillarum*. Pp. 201. Hamburg. 1820.

Lehmann, J. G. C. Revisio Potentillarum. *Nova Acta Acad. Leop.-Carol.* **23**, supplement. Pp. xiv + 230 + 64 plates. 1856.

Nestler, C. G. *Monographia de Potentilla*. Pp. 80. Paris. 1816.

Pawlowski, B. De generis Potentilla L. serie Crassinerviae (Th. Wolf) B. Pawl. nec non de taxis affinibus. *Fragm. Fl. Geobot.* **11**, 53–91. 1965.

Rhodes, H. L. J. The cultivated shrubby Potentillas. *Baileya* **2**, 89–96. 1954.

Rousi, A. Biosystematics on the species aggregate Potentilla anserina L. *Ann. Bot. Fenn.* **2**, 47–112. 1965.

Rydberg, P. A. A monograph of the North American Potentillae. *Mem. Bot. Columbia Univ.* **2**, 1–223 + 112 plates. 1898.

Trattinick, L. *Rosacearum Monographia* **4**, 1–143. Vindobonae. 1824.

Wolf, T. Monographie der Gattung Potentilla. *Bibl. Bot.* **71**, 1–714. 1908.

Zimmeter, A. Die europäischen Arten der Gattung Potentilla. *Jahresb. Staats-ober-Realschule Steyr* **1884**. Pp. 31. 1884.

Zimmeter, A. Beiträge zur Kenntnis der Gattung Potentilla. *Programm K.K. ober-Rezlschule Innsbruck* **1889**. Pp. 36. 1889.

SIBBALDIA

Muravjova, O. A. The genus Sibbaldia and its species. *Acta Inst. Bot. Acad. Sci. URSS sér.* 1 **2**, 217–240. 1936.

DUCHESNEA

Hara, H. and Kurasowa, S. On the Duchesnea indica group. *J. Jap. Bot.* **34**, 161–166. 1959.

FRAGARIA

STAUDT, G. Taxonomic studies in the genus Fragaria: typification of Fragaria species known at the time of Linnaeus. *Canad. J. Bot.* **40**, 869–886. 1962.

TRATTINICK, L. *Rosacearum Monographia* **3**, 150–169. Vindobonae. 1823.

GEUM

BOLLE, F. Ein Übersicht über die Gattung Geum L. und ihr nahestehenden Gattungen. *Fedde. Rep. Beih.* **72**, 1–119. 1933.

GAJEWSKI, W. A cytogenetic study of the genus Geum L. *Monogr. Bot.* **4**, 1–416. 1957.

SCHEUTZ, N. J. Prodromus monographiae Georum. *Nova Acta Regiae Soc. Sci. Upsal. ser.* 3. Pp. 69. 1869.

TRATTINICK, L. *Rosacearum Monographia* **3**, 110–146. Vindobonae. 1823.

DRYAS

ELKINGTON, T. T. Studies on the variation of the genus Dryas in Greenland. *Meddel. om Grønl.* **178** (1), 1–156. 1965.

HULTÉN, E. Studies in the genus Dryas. *Svensk Bot. Tidskr.* **53**, 507–542. 1959.

JUZEPCZUK, S. V. Beitrag zur Systematik der Gattung Dryas L. *Izvest. Glav. Bot. Sada* **28**, 306–327. 1928.

PORSILD, A. E. The genus Dryas in North America. *Canad. Field-Nat.* **61**, 175–192. 1947.

AGRIMONIA

MEYER, C. A. Revision des espèces du genre Agrimonia. *Ann. Sci. Nat. sér.* 2 **18**, 373–380. 1842.

MEYER, C. A. Revision der Arten der Gattung Agrimonia. *Bull. Sci. Acad. Imp. Sci. St. Pétersb.* **10**, 336–349. 1842.

SKALICKÝ, V. Ein Beitrag zur Erkenntnis der europäischen Arten der Gattung Agrimonia L. *Acta Hort. Bot. Pragensis* **1**, 87–108. 1962.

TRATTINICK, L. *Rosacearum Monographia* **4**, 157–163. Vindobonae. 1824.

WALLROTH, F. W. Monographischer Versuch über die Gewächs-Gattung Agrimonia Celsi. *Beitr. Bot. (Leipzig)* **1**, 1–61. 1842.

ALCHEMILLA

BUSER, R. Zur Kenntnis der schweizerischen Alchemilleen. *Ber. Schweiz. Bot. Ges.* **4**, 41–80. 1894.

DE LANGHE, J. E. and REICHLING, L. Les espèces d'Alchemilla du groupe vulgaris en Belgique et au Grande-Duché de Luxembourg. *Bull. Soc. Nat. Luxemb.* **59**, 133–148. 1955.

FRÖHNER, S. Zwei Alchemilla-Probleme aus Nordeuropa. *Bot. Not.* **117**, 33–56. 1964.

HADAČ, E. *Übersicht der Alchemilla-Arten Böhmens*, in Klášterský, I., *P. M. Opiz und seine Bedeutung für die Pflanzentaxonomie*, 135–158. Praha. 1958.

KLOOS, A. W. Jr. De Nederlandse vormen von Alchemilla vulgaris L. *Nederl. Kruidk. Arch.* **43**, 120–146. 1933.

LINDBERG, H. Die nordischen Alchemilla vulgaris-Formen und ihre Verbreitung. Ein Beitrag zur Kenntnis der Einwanderung den Flora Fennoscandiens mit besonderer Rücksicht auf die Finländische Flora. *Acta Soc. Sci. Fenn.* **37** (10). Pp. 172. 1909.

PAWLOWSKI, B. Przywrotniki zebrane w Czasie Trzech podrozy Balkánskich. *Acta Soc. Bot. Pol.* **22**, 245–258. 1953.

POELT, J. Die Gattung Alchemilla in Südbayern ausserhalb der Alpen. *Ber. Bayer. Bot. Ges.* **32**, 97–107. 1958.

ROTHMALER, W. Systematische Vorarbeiten zu einer Monographie der Gattung Alchemilla L. *Fedde. Rep.* **33**, 342–350. 1934: *loc. cit.* **38**, 33–43. 1935: *loc. cit.* **40**, 208–212. 1936: *loc. cit.* **42**, 111–125. 1937: *loc. cit.* **46**, 122–132. 1939: *loc. cit.* **50**, 78–80, 245–255. 1941: *loc. cit.* **66**, 192–234. 1962.

ROTHMALER, W. Systematik und Geographie der Subsektion Calycanthum der Gattung Alchemilla L. *Fedde. Rep. Beih.* **100**, 59–93. 1939.

SAMUELSSON, G. Die Verbreitung der Alchemilla-Arten aus der vulgaris-Gruppe in Nordeuropa (Fennoskandien und Dänemark). *Acta Phyt. Suec.* **16**, 1–159. 1943.

SOUGNEZ, N. and LAWALRÉE, A. Les Alchemilla de Belgique. *Bull. Jard. Bot. Bruxelles* **29**, 389–423. 1959.

TURESSON, G. Variation in the apomictic microspecies of Alchemilla vulgaris. *Bot. Not.* **1943**, 413–427. 1943: *loc. cit.* **109**, 400–404. 1956: *loc. cit.* **110**, 413–422. 1957.

WALTERS, S. M. Alchemilla vulgaris L. agg. in Britain. *Watsonia* **1**, 6–18. 1949.

APHANES

ROTHMALER, W. Systematische Vorarbeiten zu einer Monographie der Gattung Alchemilla, 3. Notizen über das Subgenus Aphanes (L.). *Fedde. Rep.* **38**, 36–38. 1935.

SANGUISORBA (includes *POTERIUM*)

NORDBORG, G. Sanguisorba L., Sarcopoterium Spach and Bencomia Webb et Berth. Delimitation and subdivision of the genera. *Opera Bot.* **11** (2), 1–103. 1966.

SPACH, E. Revisio generis Poterium. *Ann. Sci. Nat. sér.* 3 *Bot.* **5**, 34–44. 1846.

ACAENA

BITTER, G. Die Gattung Acaena. Vorstudien zu einer Monographie. *Bibl. Bot.* **74**, 1–248. 1911: *loc. cit.* **74**, 249–336. 1912.

BITTER, G. Weitere Untersuchungen über die Gattung Acaena. *Fedde. Rep.* **10**, 489–501. 1912.

GRONDONA, E. Las especies argentinas del género Acaena (Rosaceae). *Darwiniana* **13**, 209–342. 1964.

ROSA

ANDREWS, H. C. *Roses: or A Monograph of the genus Rosa.* Vol. 1. 65 plates. London. 1805: vol. 2. 64 plates. 1828.

BORBÁS, V. A Magyar birodalom vadon termó Rózsái monographiájának kisiérlete. (Primitiae monographiae Rosarum imperii Hungarici). *Math. Term. Közl.* **16**, 306–560. 1880.

BOULENGER, G. A. Les Roses d'Europe de l'herbier Crépin. *Bull. Jard. Bot. Bruxelles* **10**, 1–417. 1924–25: *loc. cit.* **12**, 1–192. 1931.

*CHRIST, H. *Die Rosen der Schweiz.* Basel. 1873.

CRÉPIN, F. Primitiae Monographiae Rosarum. *Bull. Soc. Bot. Belg.* **8**, 226–349. 1869: *loc. cit.* **11**, 15–130. 1872: *loc. cit.* **13**, 242–290. 1874: *loc. cit.* **14**, 3–46, 137–168. 1874: *loc. cit.* **15**, 12–100. 1876: *loc. cit.* **18**, 221–416. 1879: *loc. cit.* **21**, 7–196. 1882.

CRÉPIN, F. Tableau analytique des Roses européennes. *Bull. Soc. Bot. Belg.* **31**, 66–92. 1892.

DUMORTIER, B. C. J. Monographie des Roses de la Flore Belge. *Bull. Soc. Bot. Belg.* **6**, 237–297. 1867.

GANDOGER, M. *Monographia Rosarum Europae & Orientis.* Vol. 1. Pp. 338: vol. 2. Pp. 486. Paris. 1892. Vol. 3. Pp. 418: vol. 4. Pp. 601. 1893.

KELLER, R. *Rosa,* in Ascherson, P. and Graebner, P., *Syn. Mitteleur. Fl.* **6** (1), 32–384. 1900–05.

KELLER, R. Synopsis Rosarum Spontaneaurum Europae Mediae. *Denkschr. Schweiz. Naturf. Ges.* **65**. Pp. x + 796 + 40 plates. 1931.

LINDLEY, J. *Rosarum Monographia.* Pp. 156. London. 1820.

REGEL, E. A. von. *Tentamen Rosarum Monographiae.* Pp. 114. St. Pétersburg. 1877.

SERINGE, N. C. *Rosa,* in de Candolle, A. P., *Prodromus* **2**, 597–625. 1825.

TRATTINICK, L. *Rosacearum Monographia.* Vols 1 and 2. Vindobonae. 1823.

WALLROTH, F. G. *Rosa Plantarum Generis Historia Succincta.* Pp. xii + 311. Nordhusae. 1828.

WILLMOTT, E. A. *The Genus Rosa.* 2 vols (25 fasc.). Pp. xvi + xxvii + 552. London. 1910–14.

WOLLEY-DOD, A. H. A revision of the British Roses. *Supplement J. Bot.* **68** and **69**. Pp. 111. 1930–31.

WOODS, J. A synopsis of the British species of Rosa. *Trans. Linn. Soc.* **12**, 159–234. 1818.

PRUNUS

DOMIN, K. O proměnlivosti trnky (Prunus spinosa L.). *Rozpr. České Akad. Věd Umeni* **54** (27), 1–39 + 5 plates. 1945.

DOMIN, K. O původo slivy a švestky a základy botanické klasifikace těchto ovocných dřevin. *Rozpr. České Akad. Věd Umeni* **54** (28), 1–96 + 2 plates. 1945.

KALKMAN, C. The Old World species of Prunus subg. Laurocerasus including those formerly referred to Pygeum. *Blumea* **13**, 1–115. 1965.

KOEHNE, E. Die geographische Verbreitung der Kirschen, Prunus subgen. Cerasus. *Mitt. Deutsch. Dendr. Ges.* **1912**, 168–183. 1912.

MEYER, K. Kulturgeschichtliche und Systematische Beiträge zur Gattung Prunus. *Fedde. Rep. Beih.* **22**, 1–64. 1923.

COTONEASTER

BAUER, G. Beiträge zur Systematik der Gattung Cotoneaster. *Mitt. Deutsch. Dendr. Ges.* **1933**, 77–80. 1933.

KLOTZ, G. Übersicht über die in Kultur befindlichen Cotoneaster-Arten und-Formen. *Wiss. Zeitschr. Univ. Halle Math.-Nat.* **6**, 945–982. 1957.

ZABEL, H. Die Gattung der Zwergmispeln, Cotoneaster Medikus. *Mitt. Deutsch. Dendr. Ges.* **1897**, 14–32. 1897.

ZABEL, H. Nachträge zur Monographie der Gattung Cotoneaster in Mitteilungen der Deutschen Dendrologischen Gesellschaft 1897s, 14–30. *Mitt. Deutsch. Dendr. Ges.* **1898**, 37–38. 1898.

CRATAEGUS

GANDOGER, M. Révision du genre Crataegus, pour les sections des C. oxyacantha L. et oxyacanthoides Thuill. *Bull. Soc. Bot. France* **18**, 442–452. 1871.

KRUSCHKE, E. P. Contributions to the taxonomy of Crataegus. *Milwaukee Publ. Mus. Publ. Bot.* No. 3. Pp. 273. 1965.

LANGE, J. *Revisio Specierum Generis Crataegi.* Pp. iv + 106 + 10 plates. Copenhagen. 1897.

PALMER, E. J. Synopsis of North American Crataegi. *J. Arnold Arb.* **6**, 5–128. 1925.

PALMER, E. J. Crataegus in the north-eastern and central United States and adjacent Canada. *Brittonia* **5**, 471–490. 1946.

PÉNZES, A. Crataegus-Studien. *Kert. Föisk. Évkon.* **18**, 107–137. 1956.

*REGEL, E. A. von. *Revisio Specierum generis Crataego.*

AMELANCHIER

*JONES, G. N. American species of Amelanchier. *Illinois Biol. Monogr.* **20** (2). 1946.

NIELSEN, E. A taxonomic study of the genus Amelanchier in Minnesota. *Amer. Midl. Nat.* **22**, 160–206. 1939.

WIEGAND, K. M. The genus Amelanchier in eastern North America. *Rhodora* **14**, 117–161. 1912.

WIEGAND, K. M. Additional notes on Amelanchier. *Rhodora* **22**, 147–151. 1920.

SORBUS

DÜLL, R. Die Sorbus-Arten und ihre Bastarde in Bayern und Thüringen. *Ber. Bayer. Bot. Ges.* **34**, 11–65. 1961.

FRITSCH, K. Zur Systematik der Gattung Sorbus. *Österr. Bot. Zeitschr.* **48**, 1–4, 47–49, 161–171. 1898: *loc. cit.* **49**, 381–385, 426–429. 1899.

GABRIELIAN, E. The genus Sorbus L. in Turkey. *Notes Roy. Bot. Gard. Edinb.* **23**, 483–496. 1961.

HEDLUND, T. Monographie der Gattung Sorbus. *Kung. Svensk Vet.-Akad. Hand.* **35** (1), 1–147. 1901.

JONES, G. N. A synopsis of the North American species of Sorbus. *J. Arnold Arb.* **20**, 1–43. 1939.

KÁRPÁTI, Z. Bemerkungen über einige Sorbus-Arten. *Agrártud. Egyetem Kert Kar. Közl.* **12**, 119–159. 1948.

KÁRPÁTI, Z. Die Sorbus-Arten Ungarns und der angrenzenden Gebiete. *Fedde. Rep.* **62**, 71–334. 1960.

KÁRPÁTI, Z. Adatok az északi Kárpátok Sorbus-ainak ismeretéhez. *Bot. Közl.* **52**, 135–140. 1966.

KOVANDA, M. Taxonomical studies in Sorbus subg. Aria. *Acta Dendr. Čechoslov.* **3**, 23–70. 1961.

KOVANDA, M. Spontaneous hybrids of Sorbus in Czechoslovakia. *Acta Univ. Carolineae* **1961**, 41–83. 1961.

SALMON, C. E. Notes on Sorbus. *J. Bot.* **68**, 172–177. 1930.

WARBURG, E. F. *Sorbus*, in Clapham, A. R., Tutin, T. G. and Warburg, E. F. *Fl. Brit. Isles*, 539–556. Cambridge. 1952. Edition 2, 423–437. 1962.

WARBURG, E. F. Some new names in the British flora. *Watsonia* **4**, 43–46. 1957.

WILMOTT, A. J. Typification of some British Sorbi. *J. Bot.* **77**, 204–207. 1939.

MALUS

HENNING, W. Morphologische-Systematische und Genetische Untersuchungen an Arten und Artbastarden der Gattung Malus. *Zuchter* **17/18**, 289–349. 1947.

CRASSULACEAE

SEDUM

BOUVET, G. Revision des Sedum (groupe reflexum) de l'Herbier Boreau. *Rev. Bot.* **1**, 156–160. 1883.

FRÖDERSTRÖM, H. The genus Sedum L.: a systematic essay. *Meddel. Göt. Bot. Trad.* **5**, appendix, 1–75. 1930: *loc. cit.* **6**, appendix, 5–111. 1931: *loc. cit.* **7**, appendix, 3–126. 1932: *loc. cit.* **10**, appendix, 5–262. 1935.

JALAS, J. and RÖNKKO, M. T. A contribution to the cytotaxonomy of the Sedum telephium group. *Arch. Soc. Zool.-Bot. Fenn. Vanamo* **14**, 112–116. 1960.

PRAEGER, R. L. An account of the genus Sedum as found in cultivation. *J. Roy. Hort. Soc.* **46**, 1–314. 1921.

RAYMOND-HAMET, M. Contribution a l'étude phytogeographique du genre Sedum. *Candollea* **4**, 1–64. 1929.

SEMPERVIVUM

CORREVON, H. *Les Joubarbes (Semperviva)*. Pp. 134. Bruxelles. 1924.
*INGWERSEN, W. E. T. *The Genus Sempervivum.*
PRAEGER, R. L. *An Account of the Sempervivum Group*. Pp. 265. London. 1932.

UMBILICUS

UHL, C. H. Cytotaxonomic studies in the Subfamilies Crassuloideae, Kalanchoideae and Cotyledoneae of the Crassulaceae. *Amer. J. Bot.* **35**, 695–705. 1948.

SAXIFRAGACEAE

SAXIFRAGA

DON, D. A monograph of the genus Saxifraga. *Trans. Linn. Soc.* **13**, 341–452. 1821.
ENGLER, A. and IRMSCHER, E. *Saxifraga*, in Engler, H. G. A. (Ed.), *Das Pflanzenreich* **67** (IV. 117 (1)), 1–448. 1916: *loc. cit.* **69** (IV. 117 (2)), 449–709. 1919.
ENGLER, H. G. A. Naturgeschichte und Verbreitung des Genus Saxifraga L. *Linnaea* **35**, 1–124. 1866.
ENGLER, H. G. A. *De Genere Saxifraga L.* Pars 1 et 2. Pp. 62. Halle. 1866.
ENGLER, H. G. A. Index Criticus Specierum atque synonymorum generis Saxifraga L. *Verh. Zool.-Bot. Ges. Wien* **1869**, 513–556. 1869.
ENGLER, H. G. A. *Monographie der Gattung Saxifraga L.* Pp. 292. Breslau. 1872.
HAWORTH, A. H. *Saxifragarum Enumeratio*. Pp. xx + 62. Londoni. 1821.
HAYEK, A. Monographische Studien über die Gattung Saxifraga, 1. Die Sektion Porphyrion. *Denkschr. Akad. Wiss. Math.-Nat. Kl. (Wien)* **77**, 611–709. 1905.
LUIZET, D. Contribution à l'étude des Saxifrages du groupe des Dactyloides Tausch. *Bull. Soc. Bot. France* **57**, 525–534, 547–556, 595–603. 1910: *loc. cit.* **58**, 227–236, 365–372, 403–412, 637–644, 713–717. 1911: *loc. cit.* **59**, 42–51, 120–129, 148–157, 529–537, 681–685. 1912: *loc. cit.* **60**, 32–39, 58–64, 106–113, 297–304, 371–376. 1913: *loc. cit.* **62**, 145–151. 1915: *loc. cit.* **64**, 46–53, 75–83, 103–110. 1917: *loc. cit.* **65**, 83–89, 94–101, 103–116. 1918.
LUIZET, D. Additions à l'étude de quelques Saxifrages de la section des Dactyloides Tausch. *Bull. Soc. Bot. France* **60**, 409–414. 1913: *loc. cit.* **76**, 764–768. 1929.
PUGSLEY, H. W. The British Robertsonian Saxifrages. *J. Linn. Soc. Bot.* **50**, 267–289. 1936.
STERNBERG, C. Graf von. *Revisio Saxifragarum Iconibus Illustrata*. Pp. 60 + 25 plates. Ratisbonae. 1810: *Supplementum*. Pp. vi + 104 + 26 plates. 1822.

TEMESY, E. Der Formenkreis von Saxifraga stellaris Linné. *Phyton* **7**, 40–141.1957.

WEBB, D. A. A revision of the Dactyloid Saxifrages of north-western Europe. *Proc. Roy. Irish Acad.* **53**B, 207–240.1950.

WEBB, D. A. The mossy Saxifrages of the British Isles. *Watsonia* **2**, 22–29. 1951.

CHRYSOSPLENIUM

FRANCHET, A. Monographie du genre Chrysosplenium Tourn. *Nouv. Arch. Mus. (Paris) sér.* 3 **2**, 87–114 + 4 plates. 1890: *loc. cit.* **3**, 1–28 + 7 plates. 1891.

HARA, H. Synopsis of the genus Chrysosplenium L. (Saxifragaceae). *J. Fac. Sci. Tokyo Univ. (Bot.)* **7**, 1–90.1957.

GROSSULARIACEAE

RIBES

BERGER, A. A taxonomic review of currants and gooseberries. *Bull. New York State Exper. Station* **109**, 3–118.1924.

de JANCZEWSKI, E. Monographie des grosseilliers, Ribes L. *Mém. Soc. Phys. Genève* **35**, 199–517.1907.

SPACH, E. Revisio Grossulariearum. *Ann. Sci. Nat. sér.* 2 **4**, 16–31.1835.

THORY, C. A. *Monographie ou Histoire naturelle du genre Grosseillier.* Pp. xvi + 152 + 24 plates. Paris. 1829.

DROSERACEAE

DROSERA

DIELS, L. *Drosera,* in Engler, H. G. A. (Ed.), *Das Pflanzenreich* **26** (IV.112), 61–128.1906.

HAMET, M. R. Observations sur le genre Drosera. *Bull. Soc. Bot. France* **54**, 26–38, 52–76.1907.

LYTHRACEAE

KOEHNE, E. Lythraceae monographice describuntur. *Bot. Jahrb.* **1**, 142–178, 240–266.1880: *loc. cit.* **1**, 305–335, 426–458.1881: *loc. cit.* **2**, 136–176.1881: *loc. cit.* **2**, 395–429.1882: *loc. cit.* **3**, 129–155, 349–352.1882.

KOEHNE, E. Nachträge Lythraceae. *Bot. Jahrb.* **41**, 74–110.1907.

LYTHRUM (includes ***PEPLIS***)

KOEHNE, E. Lythraceae monographice describuntur. *Bot. Jahrb.* **1**, 264–266.1880: *loc. cit.* **1**, 305–333.1881.

KOEHNE, E. *Lythrum* (and *Peplis*), in Engler, H. G. A. (Ed.), *Das Pflanzenreich* **17** (IV.216), 56–78.1903.

KOEHNE, E. Nachtrage Lythraceae. *Bot. Jahrb.* **41**, 74–110.1907.

ELEAGNACEAE

HIPPOPHAE

*Darmer, G. *Der Sanddorn als Wild- und Kulturpflanze.* Leipzig. 1952.

THYMELACEAE

DAPHNE

Keissler, K. von. Die Arten der Gattung Daphne aus der Section Daphnanthes. *Bot. Jahrb.* **25**, 29–125. 1898.

Wikström, J. E. *Dissertatio botanica de Daphne.* Stockholmiae. 1820.

ONAGRACEAE

Spach, E. Synopsis Monographiae Onagearum. *Ann. Sci. Nat. sér.* 2 **4**, 161–178. 1835.

EPILOBIUM (includes ***CHAMAENERION***)

Andersen, S. Fremmede Arter af Slaegten Epilobium i Danmark. *Bot. Tidsskr.* **48**, 387–400. 1951.

Babington, C. C. On some species of Epilobium. *Trans. Proc. Bot. Soc. Edinb.* **5**, 85–103. 1857.

Haussknecht, C. *Monographie der Gattung Epilobium.* Pp. 318. Jena. 1884.

Lévéille, H. *Iconographie du genre Epilobium.* Pp. 328. Le Mans. 1910–11.

Melville, R. Epilobium pedunculare A. Cunn. and its allies. *Kew Bull.* **14**, 296–300. 1960.

Raven, P. H. The genus Epilobium in Turkey. *Notes Roy. Bot. Gard. Edinb.* **24**, 183–203.

Spach, E. Synopsis Monographiae Onagrearum. *Ann. Sci. Nat. sér.* 2 **4**, 161–178. 1835.

Trelease, W. A revision of the American species of Epilobium occurring north of Mexico. *Rep. Missouri Bot. Gard.* **2**, 69–117 + 48 plates. 1891.

OENOTHERA

de Vries, H. *Gruppenwise Artbildung unter Spezieller Berücksichtigung der Gattung Oenothera.* Pp. viii + 365 + 20 plates. Berlin. 1913.

Gates, R. R. *Taxonomy and Genetics of Oenothera: Forty years Study in the Cytology and Evolution of the Onagraceae.* Pp. 116 + 20 plates + 2 maps. The Hague. 1958.

Léveillé, H. *Monographie du genre Oenothera.* Pp. 406. Le Mans. 1902–3.

Linder, R. Les Oenothera récemment reconnus en France. *Bull. Soc. Bot. France* **104**, 515–525. 1957.

Munz, P. A. Studies in Onagraceae, 1. A revision of the subgenus Chylismia of the genus Oenothera. *Amer. J. Bot.* **15**, 223–240. 1928: 2. Revision of the North American species of subgenus Sphaerostigma, genus Oenothera. *Bot. Gaz.* **85**, 243–270. 1928: 3. A revision of the subgenera Taraxia and

Eulobus of the genus Oenothera. *Amer. J. Bot.* **16**, 246–257. 1929: 4. A revision of the subgenera Salpingia and Calyophis of the genus Oenothera. *Loc. cit.* **16**, 702–715. 1929: 5. The North American species of the subgenera Lavauxia and Megopterium of the genus Oenothera. *Loc. cit.* **17**, 358–370. 1930: 6. The subgenus Anogra of the genus Oenothera. *Loc. cit.* **18**, 309–327. 1931: 7. The subgenus Pachylophis of the genus Oenothera. *Loc. cit.* **18**, 728–738. 1931: 8. The subgenera Hartmannia and Gauropsis of the genus Oenothera. The genus Goyophytum. *Loc. cit.* **19**, 755–778. 1932: 9. The subgenus Raimmania of the genus Oenothera. *Loc. cit.* **22**, 645–663. 1935.

Raven, P. H. The systematics of Oenothera subgenus Chylismia. *Univ. Calif. Publ. Bot.* **34**, 1–122. 1962.

Renner, O. Wilde Oenotheren in Norddeutschland. *Flora* **131**, 182–226. 1937.

Renner, O. Europäische Wildearten von Oenothera. *Ber. Deutsch. Bot. Ges.* **60**, 448–466. 1942: *loc. cit.* **63**, 129–138. 1950: *Planta* **47**, 219–254. 1956.

Renner, O. Zur Kenntnis von Oenothera, 1. *Biol. Zentralbl.* **75**, 513–531. 1956.

Rostański, K. Some new taxa in the genus Oenothera L. subgenus Oenothera L. *Fragm. Fl. Geobot.* **11**, 499–523. 1965.

Rostański, K. Die Arten der Gattung Oenothera L. in Ungarn. *Acta Bot. Hung.* **12**, 337–349. 1966.

Spach E. Synopsis Monographiae Onagrearum. *Ann. Sci. Nat. sér.* 2 **4**, 161–178. 1835.

Stomps, T. J. Die niederländischen Oenothera-Arten. *Rec. Trav. Bot. Néerl.* **41**, 131–144. 1948.

Wein, K. Nordamerikanische Oenothera-Arten als Gartenpflanzen und Epökphyten in Europa während des 17 und 18 Jahrhunderts. *Beih. Bot. Centr.* **55**B, 415–543. 1936.

FUCHSIA

Munz, P. A. A revision of the genus Fuchsia (Onagraceae). *Proc. Calif. Acad. Sci. ser.* 4 **25** (1), 1–138. 1943.

CIRCAEA

Gagnepain, F. Revision du genre Circaea. *Bull. Soc. Bot. France* **63**, 39–43. 1916.

Raven, P. H. Circaea in the British Isles. *Watsonia* **5**, 262–272. 1963.

HALORAGACEAE

MYRIOPHYLLUM

Glück, H. *Biologische und morphologische Untersuchungen über Wasser- und Sumpfgewächse* Vol. 2. Pp. xvii + 256 + Taf. 6. Jena. 1906: vol. 4. Pp. viii + 746 + Taf. 8. 1924.

PEARSALL, W. H. The British species of Myriophyllum. *Rep. Bot. Soc. & E.C.* **10**, 619–621. 1934.

SCHINDLER, A. K. *Myriophyllum*, in Engler, H. G. A. (Ed.), *Das Pflanzenreich* **23** (IV. 225), 77–104. 1905.

GUNNERA

SCHNEGG, H. Beiträge zur Kenntnis der Gattung Gunnera. *Flora* **90**, 161–208. 1901.

SCHINDLER, A. K. *Gunnera*, in Engler, H. G. A. (Ed.), *Das Pflanzenreich* **23** (IV. 225), 77–104. 1905.

CALLITRICHACEAE

CALLITRICHE

FASSETT, N. C. Callitriche in the New World. *Rhodora* **53**, 137–155, 161–182, 185–194, 209–222. 1951.

GLÜCK, H. *Biologische und morphologische Untersuchungen über Wasser- und Sumpfgewächse*. Vol. 4. Pp. viii + 746 + Taf. 8. Jena. 1924.

HEGELMAIER, F. *Monographie der Gattung Callitriche*. Pp. 64 + 4 tables. Stuttgart. 1864.

HEGELMAIER, F. Zur Systematik von Callitriche. *Verh. Bot. Ver. Brandenburg* **9**, 1–41. 1867.

MASON, R. Callitriche in New Zealand and Australia. *Austral. J. Bot.* **7**, 295–327. 1959.

SAMUELSSON, G. Die Callitriche-Arten der Schweiz. *Veröff. Geobot. Inst. Rübel* **3**, 603–628. 1925.

SCHOTSMAN, H. D. A taxonomic spectrum of the section Eu-Callitriche in the Netherlands. *Acta Bot. Neerl.* **3**, 313–384. 1954.

SCHOTSMAN, H. D. Races chromosomiques chez Callitriche stagnalis Scop. et Callitriche obtusangula Le Gall. *Ber. Schweiz. Bot. Ges.* **71**, 5–16. 1961.

SCHOTSMAN, H. D. Notes on some Portuguese species of Callitriche. *Bol. Soc. Brot.* **35**, 95–128. 1961.

LORANTHACEAE

VISCUM

TUBEUF, K. F. von, NEKEL, G. and MARZELL, H. *Monographie der Mistel*. Pp. xii + 832. Berlin and München. 1923.

SANTALACEAE

THESIUM

HENDRYCH, R. Divisio generis Thesium L. cum specierum euroasiaticum respectu praecipuo. *Nov. Bot. Hort. Bot. Univ. Prag.* **1962**, 17–24. 1962.

CORNACEAE

CORNUS

KOEHNE, E. Die Sektion Microcarpium der Gattung Cornus. *Mitt. Deutsch. Dendr. Ges.* **1903**, 27–50.1903.

WANGERIN, W. *Cornus,* in Engler, H. G. A. (Ed.), *Das Pflanzenreich* **41** (IV.229), 43–92.1910.

SWIDA (THELYCRANIA)

MEYER, C. Über einige Cornus-Arten der Abtheilung Thelycrania. *Mém Acad. Sci. St. Pétersb.* (*Sci. Phys. Math.*) **1845**. Pp. 33.1845.

WANGERIN, W. *Cornus,* in Engler, H. G. A. (Ed.), *Das Pflanzenreich* **41** (IV.229, 43–92.1910.

ARALIACEAE

HEDERA

HIBBERD, S. *The Ivy, a monograph comprising the history, uses, characteristics and affinities of the plant, and a descriptive list of all the garden ivies in cultivation.* Pp. [viii] + 115. London. 1872.

LAWRENCE, G. H. M. and SCHULZE, A. E. The cultivated Hederas. *Gentes Herb.* **6**, 106–173.1942.

TOBLER, F. *Die Gattung Hedera.* Pp. iv + 151. Jena. 1912.

TOBLER, F. Die Gartenformen der Gattung Hedera. *Mitt. Deutsch. Dendr. Ges.* **1927**, 1–33.1927.

UMBELLIFERAE

CALESTANI, V. Contributo alla sistematica delle Ombellifere d'Europa. *Webbia* **1**, 89–280.1905.

CRANTZ, H. J. N. von *Classis Umbelliferarum Emendata.* Pp. 125 + Tab. 6. Lipsiae. 1767.

HEDGE, I. C. and LAMOND, J. M. A guide to the Turkish genera of Umbelliferae. *Notes Roy. Bot. Gard. Edinb.* **25**, 171–177.1964.

HOFFMANN, G. F. *Genera Plantarum Umbelliferarum.* Pp. xxx + 182. Moscow. 1814. Edition 2. Pp. xxxv + 222.1816.

PEARSALL, W. H. Notes on the Umbelliferae. *Rep. Bot. Soc. & E.C.* **10**, 850–860.1935.

SPRENGEL, C. *Species Umbelliferarum.* Pp. x + 154. Halae. 1818.

HYDROCOTYLE

RICHARD, A. *Monographie du genre Hydrocotyle.* Pp. 86 + 18 plates. Bruxelles. 1820.

SANICULA

SHAN, R. H. and CONSTANCE, L. The genus Sanicula (Umbelliferae) in the Old World and the New. *Univ. Calif. Publ. Bot.* **27**, 1–78.1951.

Wolff, H. *Sanicula*, in Engler, H. G. A. (Ed.), *Das Planzenreich* **61** (IV.228), 48–80, 278.1913.

ASTRANTIA

Grintzesco, J. Monographie du genre Astrantia. *Ann. Conserv. Bot. Gard. Genève* **13–14**, 66–194.1909–11.
Wolff, H. *Astrantia*, in Engler, H. G. A. (Ed.), *Das Pflanzenreich* **61** (IV.228), 80–92, 278.1913.

ERYNGIUM

Wolff, H. *Eryngium*, in Engler, H. G. A. (Ed.), *Das Planzenreich* **61** (IV.228), 106–271, 281–282.1913.

BUPLEURUM

Briquet, J. *Monographia des Bupleures des Alpes maritimes*. Genève. 1897.
Timbal-Lagrave, M. E. Essai Monographique sur les Bupleurum, Section Nervosa G. G. de la Flore Française. *Mém. Acad. Toulouse* **2**, 1–27.1882.
Timbal-Lagrave, M. E. Essai Monographique sur les Bupleurum, Section Marginata et Aristata de la Flore Française. *Mém. Acad. Toulouse* **2**, 29–44.1883.
Timbal-Lagrave, M. E. Essai Monographique sur des Bupleurum des Sections Perfoliata, Reticulata et Coriacia G. G. de la Flore Française. *Mém. Acad. Toulouse* **2**, 45–58.1884.
Wolff, H. *Bupleurum*, in Engler, H. G. A. (Ed.), *Das Pflanzenreich* **43** (IV.228), 36–173.1910.

TRINIA

Wolff, H. *Trinia*, in Engler, H. G. A. (Ed.), *Das Pflanzenreich* **43** (IV.228), 179–191.1910.

APIUM

Wolff, H. *Apium*, in Engler, H. G. A. (Ed.), *Das Pflanzenreich* **90** (IV.228 (2)), 26–58, 358–362.1927.

PETROSELINUM

Wolff, H. *Petroselinum*, in Engler, H. G. A. (Ed.), *Das Pflanzenreich* **90** (IV.228 (2)), 63–68, 363.1927.

SISON

Wolff, H. *Sison*, in Engler, H. G. A. (Ed.), *Das Pflanzenreich* **90** (IV.228 (2)), 71–74, 363.1927.

CICUTA

WOLFF, H. *Cicuta,* in Engler, H. G. A. (Ed.), *Das Pflanzenreich* **90** (IV.228 (2)), 75–86, 364.1927.

AMMI

WOLFF, H. *Ammi,* in Engler, H. G. A. (Ed.), *Das Pflanzenreich* **90** (IV.228 (2)), 115–123, 368.1927.

FALCARIA

WOLFF, H. *Falcaria,* in Engler, H. G. A. (Ed.), *Das Pflanzenreich* **90** (IV.228 (2)), 129–133, 368.1927.

CARUM

WOLFF, H. *Carum,* in Engler, H. G. A. (Ed.), *Das Pflanzenreich* **90** (IV.228 (2)), 143–167, 369–372.1927.

BUNIUM

WOLFF, H. *Bunium,* in Engler, H. G. A. (Ed.), *Das Pflanzenreich* **90** (IV.228 (2)), 186–212, 373–374.1927.

PIMPINELLA

WEIDE, H. Systematische Revision der Arten Pimpinella saxifraga L. und Pimpinella nigra Willd. in Mitteleuropa. *Fedde. Rep.* **64**, 240–268.1962.
WOLFF, H. *Pimpinella,* in Engler, H. G. A. (Ed.), *Das Pflanzenreich* **90** (IV. 228 (2)), 219–319, 374–376.1927.

AEGOPODIUM

WOLFF, H. *Aegopodium,* in Engler, H. G. A. (Ed.), *Das Pflanzenreich* **90** (IV. 228 (2)), 327–332.1927.

SIUM

WOLFF, H. *Sium,* in Engler, H. G. A. (Ed.), *Das Pflanzenreich* **90** (IV.228 (2)), 341–358, 376.1927.

BERULA

WOLFF, H. *Berula,* in Engler, H. G. A. (Ed.), *Das Pflanzenreich* **90** (IV.228 (2)), 336–341, 376.1927.

OENANTHE

SIMON, E. Notice sur quelques Oenanthe. *Rev. Bot. Syst. Géogr. Bot.* **1**, 65–70, 86–104.1903.

PEUCEDANUM

Boşcaiu, N. and Raţiu, F. Observaţii sistematice şi anatomice asupra unor specii de Peucedanum din secţia Peucedanum. *Contr. Bot.* (*Cluj*) **1965**, 299–312. 1965.

Castelani, V. Conspectus specierum europearum generis Peucedani. *Boll. Soc. Bot. Ital.* **1905**, 193–201. 1905.

HERACLEUM

Mandenova, I. P. *Heracleum*, in Komarov, V. L., *et al.* (Eds.), *Fl. URSS* **17**, 223–260. 1951.

Mandenova, I. P. Taxonomic review of Turkish species of Heracleum. *Notes Roy. Bot. Gard. Edinb.* **24**, 173–181. 1962.

Timbal-Lagrave, E. and Marçais, E. Essai Monographique sur les espèces françaises du genre Heracleum. *Rev. Bot.* **7**, 323–389. 1889.

DAUCUS

Nehou, J. Recherches sur la taxonomie du genre Daucus (Ombellifères) en Bretagne. *Bull. Soc. Sci. Bretagne* **36**, 81–107. 1962.

Onno, M. Die Wildformen von Daucus sect. Carota. *Beih. Bot. Centr.* **56**B, 86–136. 1937.

Thellung, A. Die Linnéschen Daucus-Arten im Lichte der Original-Herbar, exemplare. *Fedde. Rep.* **22**, 305–315. 1926.

CUCURBITACEAE

BRYONIA

Cogniaux, C. A. and Harms, H. *Bryonia*, in Engler, H. G. A. (Ed.), *Das Pflanzenreich* **88** (IV. 275 (2)), 75–93. 1924.

CITRULLUS

Cogniaux, C. A. and Harms, H. *Citrullus*, in Engler, H. G. A. (Ed.), *Das Pflanzenreich* **88** (IV. 275 (2)), 105–112. 1924.

CUCUMIS

Cogniaux, C. A. and Harms, H. *Cucumis*, in Engler, H. G. A. (Ed.), *Das Pflanzenreich* **88** (IV. 275 (2)), 116–160. 1924.

CUCURBITA

Bailey, L. H. The domesticated Cucurbitas. *Gentes Herb.* **2**, 63–115. 1929.

Bailey, L. H. Species of Cucurbita. *Gentes Herb.* **6**, 267–322. 1943.

LAGENARIA

Cogniaux, C. A. and Harms, H. *Lagenaria*, in Engler, H. G. A. (Ed.), *Das Pflanzenreich* **88** (IV. 275 (2)), 200–209. 1924.

ARISTOLOCHIACEAE

KLOTZCH, F. Die Aristolochiaceae des Berliner Herbariums. *Monatsber. Koenigl. Akad. (Berlin)* **1859**, 571–626. 1859.

ASARUM

HEMSLEY, W. B. The genus Asarum. *Gard. Chron. ser.* 3 **7**, 420–422. 1890.

KLOTZCH, F. Die Aristolochiaceae des Berliner Herbariums. *Monatsber. Koenigl. Akad. (Berlin)* **1859**, 571–626. 1859.

PRISZTER, S. Die Art Asarum europaeum L. und ihre Formenkreis. *Ann. Univ. Debrec.* **1**, 201–207. 1951.

ARISTOLOCHIA

DAVIS, P. H. and KHAN, M. S. Aristolochia in the Near East. *Notes Roy. Bot. Gard. Edinb.* **23**, 515–546. 1959.

DUCHARTRE, P. E. S. *Aristolochia*, in de Candolle, A. P., *Prodromus* **16** (1), 421–498. 1864.

KLOTZCH, F. Die Aristolochiaceae des Berliner Herbariums. *Monatsber. Koenigl. Akad. (Berlin)* **1859**, 571–626. 1859.

EUPHORBIACEAE

MERCURIALIS

PAX, F. and HOFFMAN, K. *Mercurialis*, in Engler, H. G. A. (Ed.), *Das Pflanzenreich* **63** (IV. 147 (7)), 271–282. 1914.

EUPHORBIA

*BAILLON, H. E. *Monographie des Euphorbiacées*. Paris. 1874.

BOISSIER, E. *Euphorbia*, in de Candolle, A. P., *Prodromus* **15** (2), 1–187. 1866.

CROIZAT, L. "Euphorbia esula" in North America. *Amer. Midl. Nat.* **33**, 231–243. 1945.

KHAN, M. S. Taxonomic revision of Euphorbia in Turkey. *Notes Roy. Bot. Gard. Edinb.* **25**, 71–161. 1964.

KROCHMAL, A. Seeds of weedy Euphorbia species and their identification. *Weeds* **1**, 243–255. 1952.

PROKHANOV, V. I. *Euphorbia*, in Komarov, V. L., *et al.* (Eds.), *Fl. URSS* **14**, 304–495. 1949.

ROEPER, J. *Enumeratio Euphorbiarum*. Pp. viii + 68. Gottingae. 1864.

ROSSLER, L. Vergleichinde Morphologie der Samen europäische Euphorbia-Arten. *Beih. Bot. Centr.* **62** (1), 97–174. 1943.

VINDT, J. Monographie des Euphorbiacées du Maroc, 1. Revision et Systematique. *Trav. Inst. Sci. Chérif.* **6**, xx + 217. 1953.

POLYGONACEAE

DANSER, B. H. Bijdragen tot de kennis van eenige Polygonacee. *Nederl. Kruidk. Arch.* **1920**, 218–250. 1921.

POLYGONUM

Britton, C. E. British Polygona Section Persicaria. *J. Bot.* **71**, 90–98. 1933.
Chrtek, J. Proměnlivost drůhu Polygonum aviculare L. v ČSR. *Preslia* **28**, 362–368. 1956.
Danser, B. H. Bijdragen tot de kennis van eenige Polygonaceae. *Nederl. Kruidk. Arch.* **1920**, 218–250. 1921.
Danser, B. H. Die Nederlandsche Polygonum-bastarden. *Nederl. Kruidk. Arch.* **1921**, 156–166. 1922.
Gandoger, M. Revue du genre Polygonum. *Rev. Bot.* **1**, 19–26, 41–56, 65–71, 86–98, 141–151. 1882: *loc. cit.* **1**, 169–175, 193–202. 1882.
Meisner, C. F. *Monographiae Generis Polygoni Prodromus.* Pp. iv + 117. Genevae. 1826.
Meisner, C. F. *Polygonum,* in de Candolle, A., *Prodromus* **14**, 83–143. 1856.
Mertens, T. R. and Raven, P. H. Taxonomy of Polygonum section Polygonum (Avicularia) in North America. *Madroño* **18**, 85–92. 1965.
Scholz, H. Die Systematik der europäischen Polygonum aviculare. *Ber. Deutsch. Bot. Ges.* **71**, 427–434: *loc. cit.* **72**, 63–72. 1959.
Small, J. K. A monograph of the North American species of Polygonum. *Mem. Bot. Columbia Coll.* **1**, 1–183 + 85 plates. 1895.
Steward, A. R. The Polygonaceae of eastern Asia. *Contr. Gray Herb.* **80**. Pp. 129 + 4 plates. 1930.
Styles, B. T. The taxonomy of Polygonum aviculare and its allies in Britain. *Watsonia* **5**, 177–214. 1962.
Vindt, J. Le genre Polygonum L. sect. Avicularia Meisn. au Maroc. *Bull. Soc. Sci. Nat. Maroc.* **31**, 27–36. 1952.

FAGOPYRUM

Meisner, C. F. *Fagopyrum,* in de Candolle, A., *Prodromus* **14**, 143–144. 1856.

RUMEX

Campderá, F. *Monographie des Rumex.* Pp. 169 + 3 plates. Montpellier. 1819.
Danser, B. H. Bijdrage tot de kennis van eenige Polygonaceae. *Nederl. Kruidk. Arch.* **1920**, 218–250. 1921.
Danser, B. H. Bijdrage tot de kennis der Nederlandsche Rumices. *Nederl. Kruidk. Arch.* **1921**, 167–228. 1922.
Lawalrée, A. Le genre Rumex sous-genre Acetosella en Belgique. *Bull. Jard. Bot. Bruxelles* **22**, 79–86. 1952.
Lousley, J. E. Notes on British Rumices. *Rep. Bot. Soc. & E.C.* **12**, 118–157. 1939: *loc. cit.* **12**, 547–585. 1944.
Löve, A. Études cytogénétiques des Rumex, 2. Polyploidie, géographique, systématique du Rumex subgenus Acetosella. *Bot. Not.* **1941**, 155–172. 1941.
Meisner, C. F. *Rumex,* in de Candolle, A., *Prodromus* **14**, 41–74. 1856.

MURBECK, S. Die nordeuropäischen Formen der Gattung Rumex. *Bot. Not.* **1899**, 1–42. 1899.

MURBECK, S. Zur Kenntnis der Gattung Rumex. *Bot. Not.* **1913**, 201–237. 1913.

RECHINGER, K. H. Beiträge zur Kenntnis von Rumex. *Fedde. Rep.* **26**, 177. 1929: *loc. cit.* **27**, 385–391. 1930: *loc. cit.* **29**, 246–248. 1931: *loc. cit.* **33**, 353–363. 1934: *loc. cit.* **38**, 49–55. 1935: *loc. cit.* **39**, 169–173. 1936: *loc. cit.* **40**, 294–301. 1936: *loc. cit.* **49**, 1–4. 1940: *Candollea* **11**, 229–241. 1948.

RECHINGER, K. H. Vorarbeiten zu einer Monographie der Gattung Rumex, 1. *Beih. Bot. Centr.* **49** (2), 1–132. 1932: 2. Die Arten der Subsektion Patientiae. *Fedde. Rep.* **31**, 225–283. 1933: 3. Die Süd- und ZentralAmerikanischen Arten der Gattung Rumex. *Arkiv Bot.* **26** (3), 1–58. 1933: 4. Die australischen und neuscelandischen Arten der Gattung Rumex. *Österr. Bot. Zeitschr.* **84**, 31–52. 1935: 5. The North American species of Rumex. *Publ. Field Mus. Nat. Hist. Bot. Ser.* **17** (1), 1–151. 1937: 6. Versuch einer natürlichen Gliederung des Formenkreises von Rumex bucephalophorus L. *Bot. Not.* **1939**, 485–504. 1939: 7. Rumices asiatici. *Candollea* **12**, 9–152. 1949: 8. Monograph of the genus Rumex in Africa. *Bot. Not.* **1954**, *Supplement* 3, 1–114. 1954.

RECHINGER, K. H. Die Rumex-Arten der Balkanhalbinsel. *Mitt. Thür. Bot. Ver.* **50**, 193–217. 1943.

RECHINGER, K. H. *Rumex*, in Hegi, G. *Illustr. Fl. Mitteleur.* Edition 2 **3** (2), 353–400. 1958.

RECHINGER, K. H. Notes on Rumex acetosa in the British Isles. *Watsonia* **5**, 64–66. 1961.

ROTH, F. *Die Fortpflanzungsverhältnisse in der Gattung Rumex.* Bonn. 1907.

SARKAR, M. M. Cytotaxonomic studies on Rumex Section Axillares. *Canad. J. Bot.* **36**, 947–996. 1958.

TRELEASE, W. A revision of the American species of Rumex occurring north of Mexico. *Rep. Missouri Bot. Gard.* **3**, 74–98 + 33 plates. 1892.

URTICACEAE

WEDDELL, H.-A. Revue de la Famille des Urticées. *Ann. Sci. Nat. sér.* 4 *Bot.* **1**, 173–212. 1854.

WEDDELL, H.-A. *Monographie de la Famille des Urticées.* Pp. 1–591 + 19 plates. Paris. 1856–57.

WEDDELL, H.-A. Considérations générales sur la Famille des Urticées suivie de la description des Tribus et des Genres. *Ann. Sci. Nat. Sér.* 4 *Bot.* **7**, 307–398. 1857.

PARIETARIA

PACLT, J. Über die Identät von Parietaria ramiflora Mönch mit Parietaria erecta Mertens et Koch (= P. officinalis L.). *Phyton* **4**, 46–50. 1952.

PACLT, J. Nachtrage zu meiner Parietaria-Studie. *Phyton* **5**, 242–246. 1954.

WEDDELL, H.-A. Revue de la Famille des Urticées. *Ann. Sci. Nat. sér.* 4 *Bot.* **1**, 173–212. 1854.

WEDDELL, H.-A. *Monographie de la Famille des Urticées.* Pp. 1–591 + 19 plates. Paris. 1856–57.

WEDDELL, H.-A. Considérations générales sur la Famille des Urticées suivie de la description des Tribus et des Genres. *Ann. Sci. Nat. sér.* 4 *Bot.* **7**, 307–398. 1857.

URTICA

HERMANN, F. J. The perennial species of Urtica in the United States east of the Rocky Mountains. *Amer. Midl. Nat.* **35**, 773–778. 1946.

WEDDELL, H.-A. Revue de la Famille des Urticées. *Ann. Sci. Nat. sér.* 4 *Bot.* **1**, 173–212. 1754.

WEDDELL, H.-A. *Monographie de la Famille des Urticées.* Pp. 1–591 + 19 plates. Paris. 1856–57.

WEDDELL, H.-A. Considérations générales sur la Famille des Urticées suivie de la description des Tribus et des Genres. *Ann. Sci. Nat. sér.* 4 *Bot.* **7**, 307–398. 1857.

ULMACEAE

ULMUS

BANCROFT, H. The British Elms. *J. Bot.* **75**, 337–346. 1937.

CIFERRI, R. Qualche dato per una revisione degli Olmi italiani. *Atti Ist. Bot. Pavia ser.* 5 **6**, 89–94. 1949.

JACKSON, A. B. The British Elms. *New Flora and Silva* **2**, 219–229. 1930.

MELVILLE, R. Contributions to the study of British Elms, 1. What is Goodyers Elm?, *J. Bot.* **76**, 185–192. 1938: 2. The East Anglian Elm, *loc. cit.* **77**, 138–145. 1939: 3. The Plot Elm, Ulmus Plotii Druce, *loc. cit.* **78**, 181–192. 1940.

MELVILLE, R. The application of biometrical methods to the study of elms. *Proc. Linn. Soc.* **1938–39**, 152–159. 1939.

MELVILLE, R. Typification and variation in the smooth-leaved elm, Ulmus carpinifolia Gleditsch. *J. Linn. Soc. Bot.* **53**, 83–90. 1945.

MELVILLE, R. The elms of the Dumortier Herbarium. *Bull. Jard. Bot. Bruxelles* **21**, 347–351. 1951.

SPACH, E. Revisio Ulmorum europaearum et boreali-americanum. *Ann. Sci. Nat. sér.* 2 **15**, 359–365. 1841.

TOUW, A. Een voorlopig overzicht van de Nederlandse iepen. *Jaarb. Ned. Dendr. Ver.* **22**, 57–72. 1963.

MORACEAE

FICUS

CONDIT, I. J. and ENDERUD, J. A bibliography of the fig. *Hilgardia* **25**, 1–663. 1956.

SATA, N. A monographic study of the genus Ficus from the point of view of economic botany. *Contr. Inst. Hort. Econ. Bot. Taihoku Univ.* **1–2**, 1–405: *loc. cit.* **3–4**, 1–289. 1945.

JUGLANDACEAE

JUGLANS

DODE, L. A. Contribution de l'étude du genre Juglans. *Bull. Soc. Dendr. France* **1909**, 165–215. 1909.

GRAEBNER, P. Die in Deutschland winterharten Juglandaceen. *Mitt. Deutsch. Dendr. Ges.* **1911**, 186–219. 1911.

MYRICACEAE

MYRICA

CHEVALIER, A. Monographie des Myricacées; anatomie et histologie, organographie, classification et description des espèces, distribution geographique. *Mém. Soc. Imp. Sci. Nat. Cherbourg* **32**, 85–340. 1901–2.

PLATANACEAE

PLATANUS

BRETZLER, E. Studien über die Gattung Platanus L. *Nova Acta Acad. Leop.-Carol.* **77**, 115–226. 1899.

BETULACEAE

SPACH, E. Revisio Betulacearum. *Ann. Sci. Nat. sér.* 2 **15**, 182–212. 1841.

BETULA

ASCHERSON, P. and GRAEBNER, K. O. P. P. *Betula*, in *Syn. Mitteleur. Fl.* **4**, 386–412. 1910.

BUTLER, C. B. Western American birches. *Bull. Torrey Bot. Club* **36**, 421–440. 1909.

DUGLE, J. R. A taxonomic study of western Canadian species in the genus Betula. *Canad. J. Bot.* **44**, 929–1007. 1966.

GUNNARSSON, J. G. *Monografi över Skandinaviens Betulae.* Pp. xi + 136. + 32 plates. Malmo. 1925.

JANSSON, C. A. Some species and varieties of Betula section Verrucosae in East Asia and North West America. *Meddel. Göt. Bot. Trad.* **25**, 103–156. 1962.

LINDQUIST, B. On the variation in Scandinavian Betula verrucosa Ehrh., with some notes on the Betula series Verrucosae Sukacz. *Svensk Bot. Tidskr* **41**, 45–71. 1947.

NATHO, G. Variationsbreite und Bastardbildung bei mitteleuropäischen Birkensippen. *Fedde. Rep.* **61**, 211–273. 1959.

RECHINGER, K. H. *Betula*, in Hegi, G. *Illustr. Fl. Mitteleur.* Edition 2 **3** (1), 141–173. 1957.

REGEL, A. Monographische Bearbeitung der Betulaceen. *Nouv. Mém. Soc. Nat. Moscou* **13**, 61–187. 1861. Reprinted as *Monographia Betulacearum Hucusque Cognitarum.* Pp. 129. Moscow. 1861.

SPACH, E. Revisio Betulacearum. *Ann. Sci. Nat. sér.* 2 **15**, 185–198. 1841.

JENTYS-SZAFEROWA, J. Analysis of the collective species Betula alba L. on the basis of leaf-measurements. *Bull. Acad. Polon. Sci. Lett. ser. B* 1, **1949**, 175–214. 1950: *loc. cit.* **1950**, 1–63. 1950: *loc. cit.* **1951**, 1–40. 1952.

VASSILIJEV, I. G. De genere Betula L. notulae systematicae et geographicae. *Not. Syst. (Leningrad)* **21**, 93–103. 1961.

WINKLER, H. *Betula*, in Engler, H. G. A. (Ed.), *Das Pflanzenreich* **19** (IV. 61), 56–101. 1904.

ALNUS

CALLIER, A. Alnus formen der europäischen Herbarien und Gärten. *Mitt. Deutsch. Dendr. Ges.* **1918**, 39–185. 1918.

MURAI, S. Phytotaxonomical and geobotanical studies on genus Alnus in Japan, 3. Taxonomy of the whole world species and distribution of each section. *Bull. Forest. Exp. Stat. Meguro, Tokyo* **171**, 1–107. 1964.

SPACH, E. Revisio Betulacearum. *Ann. Sci. Nat. sér.* 2 **15**, 203–209. 1841.

WINKLER, H. *Alnus*, in Engler, H. G. A. (Ed.), *Das Pflanzenreich* **19** (IV. 61), 101–134. 1904.

CORYLACEAE

CARPINUS

BERGER, W. Studien zur Systematik und Geschichte der Gattung Carpinus. *Bot. Not.* **1953**, 1–47. 1953.

WINKLER, H. *Carpinus*, in Engler, H. G. A. (Ed.), *Das Pflanzenreich* **19** (IV. 61), 24–43. 1904.

WINKLER, H. Neue Revision der Gattung Carpinus. *Bot. Jahrb.* **50** Supplement, 488–508. 1914.

CORYLUS

*BEIJERINCK, W. *Jaarb. Ned. Dendr. Ver.* **1948–49**, 67–107. 1951.

BOBROV, E. G. Histoire et systématique du genre Corylus. *Sov. Bot.* **1936**, 11–51. 1936.

WINKLER, H. *Corylus*, in Engler, H. G. A. (Ed.), *Das Pflanzenreich* **19** (IV. 61), 44–56. 1904.

FAGACEAE

CASTANEA

CAMUS, A. Les Châtaigniers. Monographie des Castanea et Castanopsis. *Encyclopédie écon. sylv.* **3**. Pp. 604. Paris. 1929: Atlas. 109 plates. 1929.

Lamy, E. *Essai monographique sur le Châtaignier*. Pp. 66. Limoges. 1860.

*Vasconcellos, J. de, Carvalho, E. and Franco, J. do Amaral. *Notul. Syst.* **15** (2), 215–225. 1956.

QUERCUS

Camus, A. Les Chênes. Monographie du genre Quercus. *Encyclopédie écon. sylv.* **6–8**. Pp. ix + 2830: Atlas 621 plates. Paris. 1934–54.

Kotschy, T. *Die Eichen europa's und des Orients'. Gesammelt, zum Theil neu entdeckt und mit Hinweisung auf ihre Culturfähigkeit für Mittel-europa beschrieben.* Viennae. 1858: Reprinted with text in French and German. Paris. 1864.

Schwarz, O. Monographie der Eichen europas und des Mittelmeergebietes. *Fedde. Rep. Sonderbeih. D*, 1–200. 1936–39.

Trelease, W. The American Oaks. *Mem. Nat. Acad. Sci. Washington* **20**, 1–225. 1924.

*Vasconcellos, J. de, Carvalho, E. and Franco, J. do Amaral. *Anais Inst. Sup. Agron. (Lisboa)* **21**, 1–135. 1954.

Viciosa, C. *Revisión del genero Quercus en España*. Pp. 194. Madrid. 1950.

Wenzig, T. Die Eichen europas, Nordafrikas und des Orients: neue bearbeitet. *Jahrb. Berlin Bot. Gart.* **4**, 179–213. 1886.

SALICACEAE

POPULUS

Cansdale, G. S. *The Black Poplars and their hybrids cultivated in Britain*. Pp. 52. Oxford. 1938.

Dode, L. A. Extraits d'une monographie inédite du genre Populus. *Mém. Soc. Hist. Nat. d'Autun.* **18**, 161–231. 1905.

Gombocz, E. A Populus-nem monográfiája. *Math. Term. Közl.* **30** (1), 1–238. 1908.

Houtzagers, G. *Het geslacht Populus in Verband met zijn beteekenis voor de houtteelt*. Pp. ix + 266. Wageningen. 1937.

*Houtzagers, G. *Handboek voor de Populierenteelt*. Wageningen. 1954.

Spach, E. Revisio Populorum. *Ann. Sci. Nat. sér.* 2 **15**, 28–33. 1841.

Wesmael, A. Monographie botanique et horticole des Peupliers cultivés en Belgique. *Féd. Soc. Hort. Belg. Bull.* **1861**, 315–354. 1862.

Wesmael, A. Monographie des Peupliers. *Mém. Soc. Sci. (Hainaut) ser.* 3 **3**, 183–253. 1869. Reprinted as *Monographie de toutes les espèces connues du genre Populus*. Mons. 1869.

SALIX

Andersson, N. J. Monographia Salicum hucusque cognitorum. *Kung. Svensk. Vet.-Akad. Hand.* **6** (1), 1–180. 1867.

Andersson, N. J. *Salix*, in de Candolle, A., *Prodromus* **16** (2), 191–323. 1868.

Basins, A. P. Ivy (Salix L.) Latviïskoý SSSR. *Trudȳ Inst. Biol. Riga* **8**, 81–123. 1959.

Buser, R. Kritische Beiträge zur Kenntnis der schweizerischen Weiden. *Ber. Schweiz. Bot. Ges.* **50**, 567–788. 1940.

Camus, A. and Camus, E. G. Classification des Saules d'Europe et Monographie des Saules de France. *J. Bot. (Paris)* **18**, 177–213, 245–296, 367–372. 1904: *loc. cit.* **19**, 1–68, 87–144. 1905: *loc. cit.* **20**, 1–116. 1906.

Floderus, B. *Salix*, in Holmberg, O. R. *Skand. Fl.* **1** (6. 1), 6–160. Stockholm. 1931.

Floderus, B. Two Linnean species of Salix and their allies. *Arkiv Bot.* **29A** (18). Pp. 54 + 2 plates. 1939.

Forbes, J. *Salictum Woburnense.* Pp. xvi + 294. London. 1829.

Görz, R. Über norddeutsche Weiden. *Fedde. Rep. Beih.* **13**, 1–127. 1922.

Hoffmann, G. F. *Historia Salicum iconibus illustrata.* Vol. 1. Pp. 78 + Tab. 1–24. Lipsiae. 1785–87: vol. 2. Pp. 12 + Tab. 25–31. 1791.

Host, N. *Salix.* Vol. 1. Pp. 34 + Tab. 105. Vindobonae. 1828.

Linton, E. A. A Monograph of the British Willows. *J. Bot.* **51** Supplement. Pp. 92. 1913.

Rechinger, K. H. Key to the British species of Salix. *Watsonia* **1**, 154–162. 1949.

Rechinger, K. H. Zur Kenntnis der europäischen Salix-Arten. *Österr. Bot. Zeitschr.* **110**, 338–341. 1963.

Rouy, M. G. Les Saules hybride européens de l'Herbier Rouy. *Rev. Bot. Syst. Géogr. Bot.* **2**, 167–181. 1904: *loc. cit.* **2**, 183–188. 1905.

Raup, H. M. The willows of boreal western America. *Contr. Gray Herb.* **185**, 3–96. 1959.

Séringe, N. C. *Essai d'une Monographie des Saules de la Suisse.* Berne. 1815.

Séringe, N. C. *Révision inédite des Saules de la Suisse.* Berne. 1824.

Smith, J. E. *Salix*, in Smith, J. E., *The English Flora* **4**, 165–233. London. 1828.

Toepffer, A. Salices Bavariae. *Ber. Bayer. Bot. Ges.* **15**, 17–233. 1915.

Wade, W. *Salices, or an essay towards a general history of Sallows, Willows and Osiers, their uses, etc.* Pp. xxii + 406 + 56. Dublin. 1811.

White, F. B. A revision of the British Willows. *J. Linn. Soc. Bot.* **27**, 333–457. 1890.

Wimmer, F. *Salices Europaeae.* Pp. xcii + 288. Breslau. 1866.

ERICACEAE

LEDUM

Tolmatchev, A. Nota de speciebus eurasiaticis generis Ledum. *Not. Syst. (Leningrad)* **15**, 197–207. 1953.

RHODODENDRON

LEACH, D. G. *Rhododendrons of the World and how to grow them.* Pp. 544 + 16 plates. London. 1962.

SLEUMER, H. Ein System der Gattung Rhododendron. *Bot. Jahrb.* **74**, 512–553. 1949.

*STEVENSON, J. B. (Ed.). *The Species of Rhododendron.* London. 1930. Edition 2. 1947.

*WILDING, E. H. *Index to the genus Rhododendron.* Stoke Poges. 1920.

PHYLLODOCE

STOKER, F. The genus Phyllodoce. *New Flora and Silva* **12**, 30–42. 1939.

ARCTOSTAPHYLOS

EASTWOOD, A. A revision of Arctostaphylos with key and descriptions. *Leafl. West. Bot.* **1**, 105–127. 1934.

ARCTOUS

KARAVAEV, M. Revisio specierum generis Arctous Niedenzu. *Not. Syst. (Leningrad)* **15**, 182–196. 1953.

CALLUNA

BEIJERINCK, W. Calluna: a monograph on the Scotch Heather. *Verh. Kon. Akad. Wetenschapp. section* 3 **38**, 1–180. 1940.

*NORDHAGEN, R. Studien über die monotypische Gattung Calluna Salisb. *Bergens Mus. Aarb.* **1937** (4). Pp. 55 + 3 plates. 1937: *loc. cit.* **1938** (1). Pp. 70. 1938.

ERICA

ANDREWS, H. C. *The Heathery: or a monograph of the genus Erica: containing coloured engravings of all known species.* 4 vols. London. 1804. Edition 2. 6 vols. 1845.

HANSEN, J. Die europäischen Arten der Gattung Erica L. *Bot. Jahrb.* **75**, 1–81. 1950.

MCCLINTOCK, D. Notes on British Heaths, 1. Species of Heather in Britain. *Heather Soc. Year Book* **1964**, 4–9. 1964: 2. Hybrids in Britain. *Loc. cit.* **1965**, 9–17. 1965.

VACCINIUM

SLEUMER, H. Vaccinideen-Studien. *Bot. Jahrb.* **71**, 375–510. 1941.

PYROLACEAE

ANDRES, H. Studie zur speziellen Systematik der Pirolaceae. *Österr. Bot. Zeitschr.* **63**, 445–450. 1913: *loc. cit.* **64**, 45–50, 232–254. 1914.

Andres, H. Beiträge zur Kenntnis der Morphologie, Phytogeographie und allgemeinen Systematik der Pirolaceae. *Verh. Bot. Ver. Brandenburg* **56**, 1–76. 1914.

PYROLA

Andres, H. Studie zur speziellen Systematik der Pirolaceae. *Österr. Bot. Zeitschr.* **63**, 445–450. 1913: *loc. cit.* **64**, 45–50, 232–254. 1914.

Andres, H. Beiträge zur Kenntnis der Morphologie, Phytogeographie und allgemeinen Systematik der Pirolaceae. *Verh. Bot. Ver. Brandenburg* **56**, 1–76. 1914.

Don, D. A monograph of the genus Pyrola. *Mem. Wern. Nat. Hist. Soc.* **5**, 220–246. 1824.

Křisa, B. Beitrag zur Gliederung der Gattung Pyrola L. in holoarktischen Gebieten. *Nov. Bot. Hort. Bot. Univ. Prag.* **1965**, 31–35. 1965.

Křisa, B. Contribution to the taxonomy of the genus Pyrola L. in North America. *Bot. Jahrb.* **85**, 612–637. 1966.

Radius, J. *Dissertatio de Pyrola et Chimophila Specimen Primium Botanicum.* Pp. 39 + Tab. 4. Lipsiae. 1821.

ORTHILIA

Andres, H. Studie zur speziellen Systematik der Pirolaceae. *Österr. Bot. Zeitschr.* **63**, 445–450. 1913: *loc. cit.* **64**, 45–50, 232–254. 1914.

Andres, H. Beiträge zur Kenntnis der Morphologie, Phytogeographie und allgemeinen Systematik der Pirolaceae. *Verh. Bot. Ver. Brandenb.* **56**, 1–76. 1914.

Andres, H. Revision der Gattung Ramischia Opiz. *Fedde. Rep.* **19**, 209–224. 1923.

Don, D. A monograph of the genus Pyrola. *Mem. Wern. Nat. Hist. Soc.* **5**, 220–246. 1824.

MONESES

Andres, H. Studie zur speziellen Systematik der Pirolaceae. *Österr. Bot. Zeitschr.* **63**, 445–450. 1913: *loc. cit.* **64**, 45–50, 232–254. 1914.

Andres, H. Beiträge zur Kenntnis der Morphologie, Phytogeographie und allgemeinen Systematik der Pirolaceae. *Verh. Bot. Ver. Brandenburg* **56**, 1–76. 1914.

Don, D. A monograph of the genus Pyrola. *Mem. Wern. Nat. Hist. Soc.* **5**, 220–246·1824.

DIAPENSIACEAE

DIAPENSIA

Diels, L. Diapensiaceen-Studien. *Bot. Jahrb.* **50** *Supplement*, 304–330. 1914.

Evans, W. E. A revision of the genus Diapensia with special reference to the Sino-Himalayan species. *Notes Roy. Bot. Gard. Edinb.* **15**, 209–236. 1927.

EMPETRACEAE

EMPETRUM

Good, R. d'O. The genus Empetrum. *J. Linn. Soc. Bot.* **47**, 489–523. 1927.

Hagerup, O. Empetrum hermaphroditum (Lge.) Hagerup. A new tetraploid, bisexual species. *Dansk Bot. Arkiv* **5**, 1–17. 1927.

PLUMBAGINACEAE

Labbe, A. Les Plumbaginacées, structure, développement, répartition, conséquences en systématique. *Trav. Lab. Biol. Vég. Grenoble-Lautaret* **1962**, 9–113. 1962.

LIMONIUM

Labbe, A. Les Plumbaginacées, structure, développement, répartition, conséquences en systématique. *Trav. Lab. Biol. Vég. Grenoble-Lautaret* **1962**, 9–113. 1962.

Pugsley, H. W. A new statice in Britain. *J. Bot.* **62**, 129–134. 1924.

Pugsley, H. W. A further new Limonium in Britain. *J. Bot.* **69**, 44–47. 1931.

Rouy, M. G. Sur quelques espèces, formes ou varietés du genre Statice. *Rev. Bot. Syst. Géogr. Bot.* **1**, 153–169. 1903: *loc. cit.* **1**, 179–186. 1904.

Salmon, C. E. Notes on Limonium. *J. Bot.* **41**, 65–74. 1903: *loc. cit.* **42**, 361–364. 1904: *loc. cit.* **43**, 5–14, 54–59. 1905: *loc. cit.* **45**, 24–26, 428–432. 1907: *loc. cit.* **46**, 1–3. 1908: *loc. cit.* **47**, 285–288. 1909: *loc. cit.* **49**, 73–77. 1911: *loc. cit.* **51**, 92–95. 1913: *loc. cit.* **53**, 237–243, 325–329. 1915: *loc. cit.* **55**, 33–34. 1917: *loc. cit.* **60**, 345–346. 1922: *loc. cit.* **61**, 97–99. 1923.

ARMERIA

Bernis, F. Revisión del género Armeria Willd., con especial referencia a los grupos ibéricos. *Anal. Inst. Bot. Cav.* **11** (2), 5–288. 1953: *loc. cit.* **12** (2), 77–252. 1954: *loc. cit.* **14**, 259–432. 1956.

Christiansen, W. Die mitteldeutschen Formenkreise der Gattung Armeria. *Bot. Archiv.* **31**, 247–265. 1930.

*Ebel, W. *De Armeriae genere Prodromus Plumbagineasum familiae. Dissertatio botanico.* Pp. iv + 44. Regimontii Prussorum, Berlin. 1840.

Labbe, A. Les Plumbaginacées, structure, développement, répartition, conséquences en systématique. *Trav. Lab. Biol. Vég. Grenoble-Lautaret* **1962**, 9–113. 1962.

Lawrence, G. H. M. Armeria, native and cultivated. *Gentes Herb.* **4**, 391–418. 1940.

Lawrence, G. H. M. The genus Armeria in North America. *Amer. Midl. Nat.* **37**, 757–779. 1947.

Szafer, W. Le genre Armeria en Pologne. *Acta Soc. Bot. Pol.* **17**, 7–28. 1946.

Wallroth, K. F. W. Monographischer Versuch über die Gewachs-Gattung Armeria Willd. *Beitr. Bot.* (*Leipzig*) 168–218. 1842.

PRIMULACEAE

PRIMULA

*Bohmig, F. *Die Gattung Primula*. Berlin. 1954.

MacWatt, J. *The Primulas of Europe*. Pp. xvi + 208. London. 1923.

Pax, F. Monographische Übersicht über die Arten der Gattung Primula. *Bot. Jahrb.* **10**, 75–241.1888.

Pax, F. and Knuth, R. *Primula*, in Engler, H. G. A. (Ed.), *Das Pflanzenreich* **22** (IV.237), 17–160, 346–347.1905.

Smith, W. W. and Fletcher, H. R. The genus Primula. *Trans. Proc. Bot. Soc. Edinb.* **33**, 122–181.1941: *loc. cit.* **33**, 209–294.1942: *J. Linn. Soc. Bot.* **52**, 321–335.1942: *Trans. Roy. Soc. Edinb.* **50**, 563–627.1942: *loc. cit.* **51**, 1–69.1943: *Trans. Proc. Bot. Soc. Edinb.* **33**, 431–487.1943: *loc. cit.* **34**, 55–158.1944: *Trans. Roy. Soc. Edinb.* **61**, 271–314.1944: *Trans. Proc. Bot. Soc. Edinb.* **34**, 402–468.1948: *loc. cit.* **35**, 180–202.1950.

Widmer, E. *Die europäischen Arten der Gattung Primula*. Pp. 154. München. 1891.

CYCLAMEN

Doorenbos, J. Taxonomy and nomenclature of Cyclamen. *Med. Landbouwhoogesch. Wageningen* **50**, 19–29.1950.

Glasau, F. Monographie der Gattung Cyclamen auf morphologisch-cytologischer Grundlage. *Planta* **30**, 507–550.1939.

Hildebrand, F. *Die Gattung Cyclamen L., eine systematische und biologische Monographie*. Pp. 190. Jena. 1898.

Pax, F. and Knuth, R. *Cyclamen*, in Engler, H. G. A. (Ed.), *Das Pflanzenreich* **22** (IV.237), 246–256, 347–348.1905.

Schwarz, O. Cyclamen-Studien. *Gartenflora* Neue Folge **1**, 11–38.1938.

Schwarz, O. Systematische Monographie der Gattung Cyclamen L. *Fedde. Rep.* **58**, 234–283.1955: *loc. cit.* **69**, 73–103.1964.

LYSIMACHIA

Klatt, F. W. Die Gattung Lysimachia L. *Abh. Nat. Ver. Hamburg* **4**, 1–45.1866.

Pax, F. and Knuth, R. *Lysimachia*, in Engler, H. G. A. (Ed.), *Das Pflanzenreich* **22** (IV.237), 256–313.1905.

TRIENTALIS

Pax, F. and Knuth, R. *Trientalis*, in Engler, H. G. A. (Ed.), *Das Pflanzenreich* **22** (IV.237), 313–316.1905.

ANAGALLIS

Clos, M. D. Les Anagallis annuels d'Europa au point de vue spécifique. *Bull. Soc. Bot. France* **44**, 292–307.1897.

MARTINOLI, G. Tassonomia ed ecologia delle specie del genero Anagallis della Sardegna. *Webbia* **15**, 1–45. 1959.
PAX, F. and KNUTH, R. *Anagallis*, in Engler, H. G. A. (Ed.), *Das Pflanzenreich* **22** (IV. 237), 321–336. 1905.
TAYLOR, P. The genus Anagallis in Tropical and South Africa. *Kew Bull.* **1955**, 321–350. 1955.

GLAUX

PAX, F. and KNUTH, R. *Glaux*, in Engler, H. G. A. (Ed.), *Das Pflanzenreich* **22** (IV. 237), 319–320. 1905.

SAMOLUS

PAX, F. and KNUTH, R. *Samolus*, in Engler, H. G. A. (Ed.), *Das Pflanzenreich* **22** (IV. 237), 336–344. 1905.

OLEACEAE

FRAXINUS

LINGELSCHEIN, A. Vorarbeiten zu einer Monographie der Gattung Fraxinus. *Bot. Jahrb.* **40**, 185–223. 1907.
LINGELSCHEIN, A. *Fraxinus*, in Engler, H. G. A. (Ed.), *Das Pflanzenreich* **72** (IV. 243(1)), 9–61. 1920.
WENZIG, T. Die Gattung Fraxinus Tourn. *Bot. Jahrb.* **14**, 165–188. 1883.
WESMAEL, A. Monographie des espèces du genre Fraxinus. *Bull. Soc. Bot. Belg.* **3**, 69–117. 1892.

LIGUSTRUM

DECAISNE, M. J. Monographie des genres Ligustrum et Syringa. *Nouv. Arch. Mus. (Paris) sér.* 2 **2**, 1–45. 1879.
HÖFKER, H. Übersicht über die Gattung Ligustrum. *Mitt. Deutsch. Dendr. Ges.* **1915**, 51–66. 1915.
HÖFKER, H. Zur Gattung Ligustrum. *Mitt. Deutsch. Dendr. Ges.* **1930**, 31–35. 1930.
KOEHNE, E. Ligustrum Sect. Ibota. *Mitt. Deutsch. Dendr. Ges.* **1904**, 68–76. 1904.
MANSFELD, R. Vorarbeiten zu einer Monographie der Gattung Ligustrum. *Bot. Jahrb.* **59**, *Beiblatt.* 132, 19–75. 1924.

APOCYNACEAE

VINCA

PICHON, M. Classification des Apocynacées, 22. Les espèces du genre Vinca. *Bull. Mus. Hist. Nat. (Paris)* **23**, 439–444. 1951.

GENTIANACEAE

GRISEBACH, A. H. R. *Genera et Species Gentianearum*. Pp. viii + 364. Stuttgartiae. 1838 (1839).

CICENDIA

GRISEBACH, A. H. R. *Genera et Species Gentianearum*, 156–160. Stuttgartiae. 1838 (1839).

CENTAURIUM

GRISEBACH, A. H. R. *Genera et Species Gentianearum*, 136–149. Stuttgartiae. 1838 (1839).

JONKER, F. P. Revisie van de Nederlandse Gentianaceae, 1. Centaurium Hill. *Nederl. Kruidk. Arch.* **57**, 168–198. 1950.

MELDERIS, A. Genetical and taxonomical studies in the genus Erythraea Rich. *Acta Hort. Bot. Univ. Latv.* **6**, 123–156. 1932.

ROBYNS, A. Essai d'étude systématique et écologique des Centaurium de Belgique. *Bull. Jard. Bot. Bruxelles* **24**, 349–398. 1954.

SCHMIDT, W. L. Einige Bemerkungen über das Genus Erythraea. *Linnaea* **7**, 467–484. 1832.

WHELDON, J. A. and SALMON, C. E. Notes on the genus Erythraea. *J. Bot.* **63**, 345–352. 1925.

BLACKSTONIA

GRISEBACH, A. H. R. *Genera et Species Gentianearum*, 116–119. Stuttgartiae. 1838 (1839).

ROBYNS, A. Le genre Blackstonia en Belgique, au Grand-Duché de Luxembourg et au Pays-Bas. *Bull. Jard. Bot. Bruxelles* **26**, 353–368. 1956.

GENTIANA

FROELICH, J. A. *De Gentiana libellus sistens specierum cognitarum descriptiones cum observationibus*. Pp. 142. Erlangae. 1796.

GRISEBACH, A. H. R. *Genera et Species Gentianearum*, 210–305. Stuttgartiae. 1838 (1839).

*KUSNEZOW, N. J. Subgenus Eu-Gentiana Kusncez. generis Gentiana Tournef. *Acta Hort. Petrop.* **15**, 1–507. 1896–1904.

GENTIANELLA

FROELICH, J. A. *De Gentiana libellus sistens specierum cognitarum descriptiones cum observationibus*. Pp. 142. Erlangae. 1796.

GRISEBACH, A. H. R. *Genera et Species Gentianearum*, 210–305. Stuttgartiae. 1838 (1839).

MURBECK, S. Studien über Gentianen aus der Gruppe Endotricha Froel. *Acta Hort. Berg.* **2** (3), 3–28. 1892.

PRITCHARD, N. M. Gentianella in Britain, 1. G. amarella, G. anglica and G. uliginosa. *Watsonia* **4**, 169–192. 1959: 2. Gentianella septentrionalis (Druce) E. F. Warb. *Loc. cit.* **4**, 218–237. 1960.

WETTSTEIN, R. von. Die europäischen Arten der Gattung Gentiana aus der Sektion Endotricha Froel. *Denkschr. Akad. Wiss. Math.-Nat. Kl. (Wien)* **64**, 309–382. 1896.

MENYANTHACEAE

NYMPHOIDES

GRISEBACH, A. H. R. *Genera et Species Gentianearum*, 336–348. Stuttgartiae. 1838 (1839).

POLEMONIACEAE

POLEMONIUM

BRAND, A. *Polemonium*, in Engler, H. G. A. (Ed.), *Das Pflanzenreich* **27** (IV. 250), 30–47. 1907.

KLOKOV, M. Polemonia Eurasiatica. *Not. Syst. (Leningrad)* **17**, 273–323. 1955.

VASSILJEV, B. H. De genre Polemonium L. notae systematicae et geographicae. *Not. Syst. (Leningrad)* **15**, 214–228. 1953.

BORAGINACEAE

INGRAM, J. Studies in the cultivated Boraginaceae, 4. A key to the genera. *Baileya* **9**, 1–12, 56. 1961.

CYNOGLOSSUM

BRAND, A. *Cynoglossum*, in Engler, H. G. A. (Ed.), *Das Pflanzenreich* **78** (IV. 252), 114–153. 1921.

OMPHALODES

BRAND, A. *Omphalodes*, in Engler, H. G. A. (Ed.), *Das Pflanzenreich* **78** (IV. 252), 96–112. 1921.

INGRAM, J. Studies in the cultivated Boraginaceae, 3. Omphalodes. *Baileya* **8**, 137–141. 1960.

ASPERUGO

BRAND, A. *Asperugo*, in Engler, H. G. A. (Ed.), *Das Pflanzenreich* **97** (IV. 253), 23–24. 1931.

SYMPHYTUM

BUCKNALL, C. A revision of the genus Symphytum Tourn. *J. Linn. Soc. Bot.* **41**, 496–551. 1913.

INGRAM, J. Studies on the cultivated Boraginaceae, 5. Symphytum. *Baileya* **9**, 92–99. 1961.

TUTIN, T. G. The genus Symphytum in Britain. *Watsonia* **3**, 280–281. 1956.

BORAGO

GUŞULEAC, M. Die monotypischen und artenarmen Gattungen der Anchusae. *Bul. Fac. Sti. Cernăuti* **2**, 394–461. 1928.

PENTAGLOTTIS

GUŞULEAC, M. Die europäischen Arten der Gattung Anchusa Linné. *Bul. Fac. Sti. Cernăuti* **1**, 73–123, 235–325. 1927.

GUŞULEAC, M. Species Anchusae generis Linn. hucusque cognitae. *Fedde. Rep.* **26**, 286–322 + tt. 79–90. 1929.

ANCHUSA

GUŞULEAC, M. Die europäischen Arten der Gattung Anchusa Linné. *Bul. Fac. Sti. Cernăuti* **1**, 73–123, 235–325. 1927.

GUŞULEAC, M. Species Anchusae generis Linn. hucusque cognitae. *Fedde. Rep.* **26**, 286–322 + tt. 79–90. 1929.

PULMONARIA

DUMORTIER, M. B. Monographie du genre Pulmonaria. *Bull. Soc. Bot. Belg.* **7**, 299–329. 1868.

KERNER, A. von. *Monographia Pulmonarium*. Pp. 53. Innsbruck. 1878.

LAWALRÉE, A. Les Pulmonaria de Belgique. *Bull. Soc. Bot. Belg.* **82**, 97–102. 1949.

*PARMENTIER, P. Contribution à l'étude du genre Pulmonaria. *Mém. Soc. Émul. Doubs*. 1891.

WILMOTT, A. J. British Pulmonarias. *J. Bot.* **55**, 223–240. 1917.

MYOSOTIS

BÉGUINOT, A. Materiali per una monografia del genere Myosotis L. *Ann. Bot. (Roma)* **1**, 275–295. 1904.

CHEVALIER, A. Les Myosotis du groupe Sylvatica et Arvensis. *Bull. Mus. Hist. Nat. (Paris) sér.* 2 **13**, 187–194. 1941.

GRAU, J. Die Zytotaxonomie der Myosotis-alpestris-und de Myosotis-silvatica-Gruppe in Europa. *Österr. Bot. Zeitschr.* **111**, 561–622. 1964.

GRAU, J. Cytotaxonomische Bearbeitung der Gattung Myosotis L., 1. Atlantische-Sippen um Myosotis secunda A. Murr. *Mitt. Bot. Staats München* **5**, 675–688. 1965.

HÜLPHERS, A. Myosotis-studier. *Svensk Bot. Tidskr.* **21**, 63–72. 1927.

STROH, G. Die Gattung Myosotis L. Versuch einer Systematischen Übersicht über die Arten. *Beih. Bot. Centr.* **61 B**, 317–346. 1941.

VERBERNE, G. Some remarks on the small-flowered Forget-me-nots. *Acta Bot. Neerl.* **8**, 330–337. 1959.

VESTERGREN, T. Systematische Beobachtungen über Myosotis silvatica (Ehrh.) Hoffm. und verwandte Formen. *Arkiv Bot.* **29A** (8). Pp. 39. 1938.

WADE, A. E. Notes on the genus Myosotis. *Rep. Bot. Soc. & E.C.* **10**, 338–392. 1933: *J. Bot.* **80**, 127–129. 1942.
WELCH, D. Water Forget-me-nots in Cambridgeshire. *Nat. Camb.* **4**, 18–27. 1961.

ALKANNA

RECHINGER, K. H. Zur Kenntnis der europäischen Arten der Gattung Alkanna. *Ann. Nat. Mus. (Wien)* **68**, 191–220. 1965.

LITHOSPERMUM

JOHNSTON, I. M. Studies in the Boraginaceae, 23. A survey of the genus Lithospermum. *J. Arnold Arb.* **33**, 299–363. 1952.
STROH, G. Vorläufiges Verzeichnis der altweltlichen Arten der Gattungen Lithospermum und Lithodora. *Beih. Bot. Centr.* **58B**, 203–212. 1938.

LAPPULA

BRAND, A. *Lappula*, in Engler, H. G. A. (Ed.), *Das Pflanzenreich* **97** (IV. 253), 136–155. 1931.

AMSINCKIA

BRAND, A. *Amsinckia*, in Engler, H. G. A. (Ed.), *Das Pflanzenreich* **97** (IV. 253), 204–217. 1931.
MACBRIDE, J. F. A revision of the North American species of Amsinckia. *Contr. Gray Herb. new ser.* **49**, 1–16. 1917.
RAY, P. M. and HARU, F. C. Studies on Amsinckia, 1. A synopsis of the genera with a study of heterostyly in it. *Amer. J. Bot.* **44**, 529–536. 1957.
SUKSDORF, W. N. Untersuchungen in der Gattung Amsinckia. *Werdenda* **1**, 47–113. 1931.

MERTENSIA

WILLIAMS, L. O. A monograph of the genus Mertensia in North America. *Ann. Missouri Bot. Gard.* **24**, 17–159. 1937.

ECHIUM

de COINCY, A.-M. Révision des espèces critiques du genre Echium. *J. Bot. (Paris)* **14**, 297–304, 322–330. 1900: *loc. cit.* **15**, 311–329. 1901: *loc. cit.* **16**, 66–68, 107–112. 1902.
de COINCY, A.-M. Énumération des Echium de la flora Atlantique. *J. Bot. (Paris)* **16**, 213–220, 226–233, 257–266. 1902.
de COINCY, A.-M. Les Echium de la section des Pachylepis sect. nov. *Bull. Herb. Boiss. sér.* 2 **3**, 261–277, 488–499. 1903.

*Klotz, G. *Die Systematische Gliederung der Gattung Echium L. Habilitationschrift.* Halle (*ined.*). 1959.

Lacaita, C. C. Revision of some critical species of Echium. *J. Linn. Soc. Bot.* **44**, 363–438. 1919.

CONVOLVULACEAE

DICHONDRA

Tharp, B. C. and Johnston, M. C. Recharacterization of Dichondra (Convolvulaceae) and a revision of the North American species. *Brittonia* **13**, 346–360. 1961.

CONVOLVULUS

Hallier, H. Bausteine zu einer Monographie der Convolvulaceen. *Bull. Herb. Boiss.* **5**, 366–387, 736–754, 804–820, 996–1013, 1021–1052. 1897: *loc. cit.* **6**, 714–724. 1898: *loc. cit.* **7**, 408–418. 1899.

*Karamanoglu, K. The species of Convolvulus in Turkey. *Comm. Fac. Sci. Univ. Ankara (Sci. Nat.)* **9**, 225–251. 1964.

IPOMOEA

House, H. D. The North American species of the genus Ipomoea. *Ann. New York Acad. Sci.* **18**, 181–263. 1908.

Matuda, E. El género Ipomoea en México. *Anal. Inst. Biol.* (*Mexico*) **34**, 85–145. 1964: *loc. cit.* **35**, 45–47. 1965: *loc. cit.* **36**, 83–106. 1966.

CALYSTEGIA

Brummitt, R. K. and Heywood, V. H. Pink-flowered Calystegiae of the Calystegia sepium complex in the British Isles. *Proc. Bot. Soc. Brit. Isles* **3**, 384–388. 1960.

CUSCUTA

Babington, C. C. On some species of Cuscuta. *Trans. Proc. Bot. Soc. Edinb.* **2**, 199–208. 1846.

Choisy, J.-D. De Convolvulaceis dissertatio tertia, complectens Cuscuta hucusque cognitorum methodicam enumerationen et descriptionen. *Mém. Soc. Phys. Nat. Genève* **9**, 262–288. 1841.

Engelmann, G. Systematic arrangement of the species of the genus Cuscuta. *Trans. Acad. Sci. St. Louis* **1**, 453–523. 1859.

Feinbrun, N. and Taub, S. The Cuscuta species of Palestine. *Israel J. Bot.* **13**, 1–23. 1964.

Gaertner, E. E. Studies of seed germination, seed identification and host relationship in Dodders, Cuscuta spp. *Mem. Cornell Univ. Agr. Exp. Stat.* **294**, 1–56. 1950.

YUNCKER, T. G. Revision of the North American and West Indian species of Cuscuta. *Illinois Biol. Monogr.* **6** (2), 1–141. 1921.

YUNCKER, T. G. The genus Cuscuta. *Mem. Torrey Bot. Club* **18**, 109–331. 1932.

SOLANACEAE

LYCIUM

FEINBRUN, N. and STEARN, W. T. Typification of Lycium barbarum L., L. afrum L. and L. europaeum L. *Israel J. Bot.* **12**, 114–123. 1964.

HITCHCOCK, C. L. A monographic study of the genus Lycium of the western hemisphere. *Ann. Missouri Bot. Gard.* **19**, 179–374. 1922.

POJARKOVA, A. Species generis Lycium L. fructibus rubris ex Asia Media et China. *Not. Syst. (Leningrad)* **13**, 238–278. 1950.

WEITZ, R. *Les Lycium européens et exotiques.* Pp. 202. Paris. 1921.

ATROPA

PASCHER, A. Über Atropa. *Flora* **148**, 84–109. 1960.

NICOTIANA

COMES, A. *Monographie du genre Nicotiana.* Pp. 80. Naples. 1899.

GOODSPEED, T. H. The genus Nicotiana. *Chron. Bot.* **16**. Pp. xxii + 536. (1954). 1955.

PHYSALIS

RYDBERG, P. A. The North American species of Physalis and related genera. *Mem. Torrey Bot. Club* **4**, 297–372. 1896.

WATERFALL, U. T. A taxonomic study of the genus Physalis in North America north of Mexico. *Rhodora* **60**, 107–114, 128–142, 152–173. 1958.

SOLANUM

BITTER, G. Solana nova vel minus cognita. *Fedde. Rep.* **10**, 529–565. 1912: *loc. cit.* **11**, 1–18, 202–237, 241–260, 349–394, 431–473. 1912: *loc. cit.* **11**, 481–491, 561–566. 1913: *loc. cit.* **12**, 1–10, 49–90, 136–162, 433–467, 542–555. 1913: *loc. cit.* **13**, 88–103, 169–173. 1914: *loc. cit.* **15**, 93–98. 1918: *loc. cit.* **16**, 10–15, 79–103. 1919: *loc. cit.* **16**, 388–409. 1920: *loc. cit.* **18**, 49–71, 301–309. 1922.

DUNAL, M. F. *Histoire naturelle, médicale et économique des Solanum.* Pp. 248. Paris. 1813.

LAWRENCE, G. H. M. The cultivated species of Solanum. *Baileya* **8**, 21–35. 1960.

SEITHE, A. Die Haararten der Gattung Solanum L. und ihre taxonomische Verwertung. *Bot. Jahrb.* **81**, 261–335. 1962.

STEBBINS, G. L. and PADDODE, E. F. The Solanum nigrum complex in Pacific North America. *Madroño* **10**, 70–81. 1949.

WESSELY, I. Die mitteleuropäischen Sippen der Gattung Solanum Sektion Morella. *Fedde. Rep.* **63**, 290–321. 1960.

CAPSICUM

*FINGERHUTH, A. *Monographia generis Capsici.* Düsseldorphii. 1832.

IRISH, H. C. A revision of the genus Capsicum with special reference to the garden varieties. *Ann. Rep. Missouri Bot. Gard.* **9**, 53–110. 1898.

TERPÓ, A. Kritische Revision der wildwachsenden Arten und der kultivierten Sorten der Gattung Capsicum L. *Fedde. Rep.* **72**, 155–191. 1966.

DATURA

*BERNHARDI, J. J. Die Arten der Gattung Datura. *Trommsd. N.J. Pharm.* **26**, 118–158. 1833.

BLAKESLEE, A. F., AVERY, A. G., SATINA, S. and RIETSAMA, J. *The Genus Datura.* Pp. xli + 289. New York. 1959.

DANERT, S. Ein Beitrag zur Kenntnis der Gattung Datura L. *Fedde. Rep.* **57**, 231–242. 1955.

LUNDSTROM, E. Datura. *Acta Hort. Berg.* **5**, 84–97. 1914.

SAFFORD, W. E. Synopsis of the genus Datura. *J. Washington Acad. Sci.* **11**, 173–189. 1921.

SAFFORD, W. E. Daturas of the Old World and the New. *Ann. Rep. Smith. Inst.* **1920**, 537–567. 1922.

SCROPHULARIACEAE

BENTHAM, G. Caractères des tribus études genres de la famille des Scrophalarinees. *Ann. Sci. Nat. sér.* 2 **4**, 178–188. 1835.

VERBASCUM

FRANCHET, M. A. Essai sur les espèces du genre Verbascum. *Mém. Soc. Acad. Maine Loire* **22**, 65–204. 1868.

FRANCHET, M. A. *Études sur les Verbascum de la France et de l'Europe centrale.* Pp. 131. Vendome. 1875.

HUBER-MORATH, A. Verbreitung der Gattungen Verbascum, Celsia und Staurophragma im Orient. *Bauhinia* **1**, 1–83. 1955.

HUBER-MORATH, A. Novitiae Florae Anatolicae, 5. Gattung Verbascum. *Bauhinia* **1**, 335–349. 1960.

MURBECK, S. Monographie der Gattung Verbascum. *Lunds Univ. Årsskr. new ser.* **29** (2), 1–630. 1933.

MURBECK, S. Nachträge zur Monographie der Gattung Verbascum. *Lunds Univ. Årsskr. new ser.* **32** (1), 1–46. 1936.

MURBECK, S. Weitere Studien über die Gattung Verbascum und Celsia. *Lunds Univ. Årsskr. new ser.* **35**, 1–72. 1939.

Schrader, H. A. *Monographia Generis Verbasci.* Sectio 1. Pp. 40. Gottingae. 1813: Sectio 2. Pp. 57. 1823.

ANTIRRHINUM

Chavannes, E. *Monographie des Antirrhinées.* Pp. 190 + 11 plates. Paris. 1833.

de Wolf, G. P., Jnr. Notes on Cultivated Scrophulariaceae, 2. Antirrhinum and Asarina. *Baileya* **4**, 55–68. 1956.

Rothmaler, W. Taxonomische Monographie der Gattung Antirrhinum. *Fedde. Rep. Beih.* **136**, 1–124. 1956.

ASARINA

de Wolf, G. P., Jnr. Notes on cultivated Scrophulariaceae, 2. Antirrhinum and Asarina. *Baileya* **4**, 55–68. 1956.

LINARIA

de Wolf, G. P., Jnr. Notes on cultivated Scrophulariaceae, 3. Linaria. *Baileya* **4**, 102–114. 1956.

CHAENORHINUM

España, T. M. Losa. Especies Españolas del género Chaenorhinum Lge. *Anal. Inst. Bot. Cav.* **21**, 543–566. 1963.

CYMBALARIA

Chevalier, A. Les espèces élémentaires françaises du genre Cymbalaria. *Bull. Soc. Bot. France* **83**, 638–653. 1936.

Cufodontis, G. Revisione monografica delle Linaria appartenenti alla seg. Cymbalaria Cham. *Arch. Bot.* **12**, 54–81, 135–158, 233–254. 1936.

Cufodontis, G. Die Gattung Cymbalaria Hill. Nachtrage und Zusammenfassung. *Bot. Not.* **1947**, 135–156. 1947.

SCROPHULARIA

Stiefelhagen, H. Systematische und pflanzengeographische Studien zur Kenntnis der Gattung Scrophularia. Vorarbeiten zu einer Monographie. *Bot. Jahrb.* **44**, 406–496. 1910.

Wydler, H. Essai monographique sur le genre Scrofularia. *Mém. Soc. Phys. Genève* **4**, 122–169. 1828.

MIMULUS

Campbell, G. R. Mimulus guttatus and related species. *Aliso* **2**, 319–335. 1950.

Grant, A. L. A monograph of the genus Mimulus. *Ann. Missouri Bot. Gard.* **2**, 99–388. 1924.

LIMOSELLA

GLÜCK, H. Limosella-Studien. Beiträge zur Systematik, Morphologie and Biologie der Gattung Limosella. *Bot. Jahrb.* **66**, 488–566. 1934.

SIBTHORPIA

HEDBERG, O. A taxonomic revision of the genus Sibthorpia. *Bot. Not.* **108**, 161–183. 1955.

DIGITALIS

HAASE-BESSELL, G. Digitalis-Studien, 1. *Zeitschr. Indukt. Abst.-Vererb.* **16**, 293–314. 1916.

IVANINA, L. I. Rod Digitalis L. (naperstianka) i yego prakticheskiye primeneniya. *Acta Inst. Bot. Acad. Sci. URSS ser.* 1 **11**, 198–308. 1955.

LINDLEY, J. *Digitalium monographia.* Pp. 27. London. 1821.

WERNER, K. Zur Nomenklatur und Taxonomie von Digitalis L. *Bot. Jahrb.* **79**, 218–254. 1960.

WERNER, K. Die Kultivierten Digitalis-Arten. *Kulturpflanze Beih.* **3**, 167–182. 1962.

WERNER, K. Die Verbreitung der Digitalis-Arten. *Wiss. Zeitschr. Univ Halle Math.-Nat. Reihe* **H6**, 453–486. 1964.

VERONICA

DE WOLF, G. P. Jnr. Notes on cultivated Scrophulariaceae, 4. Veronica. *Baileya* **4**, 143–159. 1956.

DRABBLE, E. H. and LITTLE, J. E. The British Veronicas of the Agrestis group. *J. Bot.* **69**, 180–185, 201–205. 1931.

HÄRLE, A. Die Arten und Formen der Veronica-Sektion Pseudolysimachia Koch auf Grund systematischer und experimenteller Untersuchungen. *Bibl. Bot.* **104**. Pp. vi + 86 + map + 13 plates + 14 tables. 1932.

KOCH, C. H. E. *Dissertatio inauguralis Monographia generis Veronicae.* Pp. 36. Wirceburgii. 1838.

LEHMANN, E. Geschichte und Geographie der Veronica-Gruppe Agrestis. *Bull. Herb. Boiss. sér.* 2 **8**, 229–244, 337–352, 410–425, 644–660. 1908.

LEHMANN, E. Über Zwischenrassen in der Veronica-Gruppe Agrestis. *Zeitschr. Indukt. Abst.-Vererb.* **2**, 145–208. 1909.

LEHMANN, E. Geschichte und Geographie der Veronica-Gruppe Megasperma. *Bibl. Bot.* **99**. Pp. 55 + Taf. 1. 1929.

LEHMANN, E. Polyploidie und geographische Verbreitung der Arten der Gattung Veronica. *Jahrb. Wiss. Bot.* **81**, 461–542. 1940.

LEHMANN, E. and LOHNER-SCHMITZ, M. Entwicklung und Polyploidie in der Veronica-Gruppe Agrestis. *Zeitschr. Indukt. Abst.-Vererb.* **86**, 1–34. 1954.

PENNELL, F. W. Veronica in North and South America. *Rhodora* **23**, 1–42. 1921.

RIEK, R. Systematische und pflanzengeographische Untersuchungen in der Veronica-Sektion Chamaedrys Griseb. *Fedde. Rep. Beih.* **79**, 1–68. 1935.

RÖMPP, H. Die Verwandtschaftsverhältnisse in der Gattung Veronica. *Fedde. Rep. Beih.* **50**, 1–172. 1928.

SCHLENKER, G. Systematische Untersuchungen über die Sektion Beccabunga der Gattung Veronica. *Fedde. Rep. Beih.* **90**, 1–40. 1936.

SMITH, J. E. Remarks on the genus Veronica. *Trans. Linn. Soc.* **1791**, 189–195. 1791.

PEDICULARIS

BONATI, G. *Le genre Pedicularis L.: morphologie, classification, distribution-géographique, evolution et hybridation.* Pp. viii + 168. Nancy. 1918.

LIMPRICHT, W. Studien über die Gattung Pedicularis. *Fedde. Rep.* **20**, 161–265. 1924.

STEININGER, H. Beschreibung der europäischen Arten des Genus Pedicularis. *Bot. Centralbl.* **28**, 215–219, 246–249, 279–282, 313–315, 341–342, 375–377, 388–391. 1885: *loc. cit.* **29** 23–24, 56–58, 85–89, 122–123, 154–157, 185–188, 216–221, 246–250, 278–280, 314–317, 346–349, 375–378. 1886: *loc. cit.* **29**, 25–28, 56–62, 87–93. 1887.

RHINANTHUS

CHABERT, A. Étude sur le genre Rhinanthus L. *Bull. Herb. Boiss.* **7**, 497–517. 1899.

HAMBLER, D. J. Some taxonomic investigations on the genus Rhinanthus. *Watsonia* **4**, 101–116. 1958.

SOÓ, R. von. Die Mittel- und Südosteuropäischen Arten und Formen der Gattung Rhinanthus und ihre Verbreitung in Südosteuropa. *Fedde. Rep.* **26**, 179–219. 1929.

STERNECK, J. von. Monographie der Gattung Alectorolophus. *Abh. Zool.-Bot. Ges. Wien* **1** (2), 1–150. 1901.

MELAMPYRUM

BEAUVERD, G. Monographie du genre Melampyrum L. *Mém. Soc. Phys. Genève* **38**, 219–657. 1916.

BRITTON, C. E. The genus Melampyrum in Britain. *Trans. Proc. Bot. Soc. Edinb.* **33**, 357–379. 1943.

JASIEWICZ, A. Polskie gatunki rodzaju Melampyrum L. *Fragm. Fl. Geobot.* **3**, 17–120. 1958.

SOÓ, R. von. Systematische Monographie der Gattung Melampyrum. *Fedde. Rep.* **23**, 159–176. 1926: *loc. cit.* **23**, 385–397: *loc. cit.* **24**, 127–193. 1927.

EUPHRASIA

CALLEN, E. O. Studies in the genus Euphrasia. *J. Bot.* **78**, 213–218.1940: *loc. cit.* **79**, 11–13.1941: *Rhodora* **54**, 154–156.1952.

PUGSLEY, H. W. Notes on British Euphrasias. *J. Bot.* **57**, 169–175.1919: *loc. cit.* **60**, 1–5.1922: *loc. cit.* **71**, 83–90.1933: *loc. cit.* **74**, 71–75.1936: *loc. cit.* **78**, 11–13, 89–92.1940.

PUGSLEY, H. W. New British species of Euphrasia. *J. Bot.* **67**, 224–225.1929.

PUGSLEY, H. W. A revision of the British Euphrasiae. *J. Linn. Soc. Bot.* **48**, 467–544.1930.

SMEJKAL, M. Zur Taxonomie einiger tschechoslowakischen Euphrasia-Arten. *Čas. Slez. Mus. Opavaě ser. A* **7**, 69–79.1958.

TOWNSEND, F. Monograph of the British species of Euphrasia. *J. Bot.* **35**, 321–336, 395–406, 417–426, 465–477.1897.

WETTSTEIN, R. von. *Monographie der Gattung Euphrasia.* Pp. 316. Leipzig. 1896.

ODONTITES

HOFFMANN, J. Beitrag zur Kenntnis der Gattung Odontites. *Österr. Bot. Zeitschr.* **47**, 113–117, 184–187, 233–239, 345–349.1897.

MARKLUND, G. Die Gattung Odontites in Finnland. *Acta Soc. Fauna Fl. Fenn.* **72** (16), 1–18.1955.

SCHNEIDER, U. Die Sippen der Gattung Odontites in Norddeutschland. *Fedde. Rep.* **69**, 180–195.1964.

WAISBECKER, A. Beiträge zur Kenntnis der Gattung Odontites. *Österr. Bot. Zeitschr.* **49**, 437–442.1899.

OROBANCHACEAE

LATHRAEA

BECK-MANNAGETTA, G. *Lathraea*, in Engler, H. G. A. (Ed.), *Das Pflanzenreich* **96** (IV.261), 317–326.1930.

HEINRICHER, E. *Monographie der Gattung Lathraea.* Pp. iv + 152. Jena. 1931.

OROBANCHE

ACHEY, D. M. A revision of the section Gymnocaulis of the genus Orobanche. *Bull. Torrey Bot. Club* **60**, 441–451.1933.

BECK-MANNAGETTA, G. Monographie der Gattung Orobanche. *Bibl. Bot.* **4** (19), 1–275.1890.

BECK-MANNAGETTA, G. *Orobanche*, in Engler, H. G. A. (Ed.), *Das Pflanzenreich* **96** (IV.261), 44–304.1930.

GUIMARÃES, J. D. Monographia das Orobanchaceas Portuguezas. *Broteria* **3**, 5–208.1904.

KOCH, M. Description des Orobanches de la Flora d'Allemagne. *Ann. Sci. Nat. sér.* 2 **5**, 34–48, 82–98, 146–156.1836.

PUGSLEY, H. W. Notes on Orobanche L. *J. Bot.* **78**, 105–116.1940.

VAUCHER, J. P. *Monographie des Orobanches.* Pp. ii + 72. Genève. 1827.

LENTIBULARIACEAE

PINGUICULA

CASPER, S. J. Revision der Gattung Pinguicula in Eurasien. *Fedde. Rep.* **66**, 1–148. 1962.

CASPER, S. J. Monographie der Gattung Pinguicula L. *Bibl. Bot.* **127/128**, 1–209 + Taf. 1–16. 1966.

ERNST, A. Revision der Gattung Pinguicula. *Bot. Jahrb.* **80**, 146–194. 1961.

SCHINDLER, J. Studien über einige mittel- und südeuropäische Arten der Gattung Pinguicula. *Österr. Bot. Zeitschr.* **57**, 409–421, 458–469. 1907: *loc. cit.* **58**, 13–18, 61–69. 1908.

UTRICULARIA

GLÜCK, H. *Untersuchungen über die Mitteleuropäischen Utricularia-Arten, über die Turionenbildung bei Wasserpflanzen, sowie über Ceratophyllum: Biologische und morphologische Untersuchungen über Wasser- und Sumpfgewächse.* Vol. 2. Pp. xvii + 256 + Taf. 6. Jena. 1906.

GLÜCK, H. Contributions to our knowledge of the species of Utricularia in Great Britain, with special regard to the morphology and distribution of Utricularia ochroleuca. *Ann. Bot.* **27**, 607–619. 1913.

HALL, P. M. The British species of Utricularia. *Rep. Bot. Soc. & E.C.* **12**, 100–117. 1939.

MEISTER, F. Beiträge zur Kenntnis der europäischen Arten von Utricularia. *Mém. Herb. Boiss.* **1** (12), 1–40. 1900.

ROSSBACH, G. B. Aquatic Utricularias. *Rhodora* **41**, 113–128. 1939.

VERBENACEAE

VERBENA

MOLDENKE, H. Materials towards a monograph of the genus Verbena. *Phytologia* **8**, 95–104, 108–152. 1961: *loc. cit.* **8**, 175–216, 230–272, 274–322, 371–384, 395, 453. 1962: *loc. cit.* **8**, 460–496. 1963: *loc. cit.* **9**, 8–54, 59–97, 113–181, 189–238, 267–336. 1963: *loc. cit.* **9**, 351–407, 459–480, 501–505. 1964: *loc. cit.* **10**, 56–161, 173–236, 271–319, 406–416, 490–504. 1964: *loc. cit.* **11**, 1–68, 80–142, 155–213. 1964: *loc. cit.* **11**, 219–287, 290–357, 400–422, 435–507. 1965.

PERRY, L. M. A revision of the North American species of Verbena. *Ann. Missouri Bot. Gard.* **20**, 239–362. 1933.

LABIATAE

BENTHAM, G. *Labiatarum Genera et Species, or a description of the genera and species of plants of the order Labiatae.* Vol. 1. Pp. 1–60. London. 1832: Pp. 61–323. 1833. Vol. 2. Pp. 324–566. 1834. Vol. 3. Pp. 567–645. 1834: Pp. 646–783 + iii–lxviii. 1835.

Briquet, J. Fragmenta Monographiae Labiatarum. *Bull. Soc. Bot. Genève* **5**, 20–122. 1889: *Bull. Herb. Boiss.* **2**, 119–141. 1894: *loc. cit.* **2**, 689–724. 1896: *loc. cit.* **4**, 676–696, 762–808, 847–878. 1896: *Ann. Conserv. Jard. Bot. Genève* **2**, 102–251. 1898.

MENTHA

Bentham, G. *Labiatarum Genera et Species* **1**, 168–184. 1833: *loc. cit.* **3**, 714–716. 1835.

Braun, H. Über einige Arten und Formen der Gattung Mentha. *Verh. Zool.-Bot. Ges. Wien* **40**, 351–507, 1890.

Briquet, J. Fragmenta Monographiae Labiatarum. *Bull. Soc. Bot. Genève* **5**, 34–107. 1889: *Bull. Herb. Boiss.* **2**, 691–709. 1896: *loc. cit.* **4**, 676–696, 762–784. 1896.

Déséglisé, A. and Durand, T. *Descriptions de nouvelles Menthes.* Genève. 1879.

de Wolf, G. P. Notes on cultivated Labiates, 2. Mentha. *Baileya* **2**, 3–11. 1954.

Fraser, J. Menthae Briquetianae. *Rep. Bot. Soc. & E.C.* **7**, 613–628. 1925.

Fraser, J. Menthae Britannicae. *Rep. Bot. Soc. & E.C.* **8**, 213–247. 1927.

Graham, R. A. Mint notes. *Watsonia* **1**, 88–90. 1949: *loc. cit.* **1**, 276–278. 1950: *loc. cit.* **2**, 30–35. 1951: *loc. cit.* **3**, 109–121. 1954: *loc. cit.* **4**, 70–76, 119–121. 1958.

Hylander, N. Släktet Mentha i det nordiska floraområdet. *Bot. Not.* **118**, 225–241. 1965.

Marklund, G. Mentha gentilis-Komplexet och M. dalmatica i Östfennoskandien. *Mem. Soc. Fauna Fl. Fenn.* **38**, 2–18. 1963.

Smith, J. E. Observations on the British species of Mentha. *Trans. Linn. Soc.* **5**, 171–217. 1800.

Sole, W. *Menthae Britannicae.* Pp. 55. Bath. 1798.

Topitz, A. Beiträge zur Kenntnis der Menthenflora von Mitteleuropa. *Bot. Centralbl. Beih.* **30** (2), 138–264. 1913.

LYCOPUS

Bentham, G. *Labiatarum Genera et Species* **1**, 184–188. 1833: *loc. cit.* **3**, 716. 1835.

Henderson, N. C. A taxonomic revision of the genus Lycopus. *Amer. Midl. Nat.* **68**, 95–138. 1962.

ORIGANUM

Bentham, G. *Labiatarum Genera et Species* 2, 334–339. 1834: *loc. cit.* **3**, 728. 1835.

de Wolf, G. P. Notes on cultivated Labiates, 3. Origanum and its relatives. *Baileya* **2**, 57–66. 1954.

THYMUS

Bentham, G. *Labiatarum Genera et Species* 1, 340–351. 1834: *loc. cit.* **3**, 728–729. 1835.

DOMIN, K. and JACKSON, A. B. The British species of Thymus. *J. Bot.* **46**, 33–37. 1908.

JALAS, J. Zur Systematik und Verbreitung der Fennoskandischen Formen der Kollektivart Thymus serpyllum L. em. Fr. *Acta Bot. Fenn.* **39**, 3–85. 1947.

MACHULE, M. Die mitteleuropäischen Thymus-Arten, Formen und Bastarde. *Mitt. Thür. Bot. Ges.* **1**, 13–89. 1957.

MACHULE, M. Die mitteleuropäischen Thymus-Arten, Formen und Bastarde, Nachtrag. *Mitt. Thür. Bot. Ges.* **11**, 176–207. 1960.

PIGOTT, C. D. Species delimitation and racial divergence in British Thymus. *New Phyt.* **53**, 470–495. 1954.

PIGOTT, C. D. Thymus (Biological Flora). *J. Ecol.* **43**, 365–387. 1955.

RECHINGER, K. H. Die Gattung Thymus in Persien und angrenzenden Gebieten. *Phyton* **5**, 280–303. 1954.

RONNIGER, K. Beiträge zur Kenntnis der Gattung Thymus, 1. Die britischen-Arten und Formen. *Fedde. Rep.* **20**, 321–332. 1924.

RONNIGER, K. The genus Thymus. *Rep. Bot. Soc. & E.C.* **7**, 226–239. 1924.

RONNIGER, K. Beiträge zur Kenntnis der Thymus-Flora der Balkanhalbinsel. *Fedde. Rep.* **20**, 334–336, 385–390. 1924.

RONNIGER, K. *Thymus*, in Hayek, E., Prodromus Flora Peninsulae Balcanicae 2. *Fedde. Rep. Beih.* **30** (2), 337–382. 1930. Reprinted as *Die Thymus-Arten der Balkan Halbinsel (Species balcanicae generis Thymi)*. 1930.

RONNIGER, K. Bestimmungstabelle für die Thymus-Arten des Deutsches Reiches. *Ber. Bayer. Bot. Ges.* **30**, 103–108. 1954.

ROUSSINE, N. À propos de Thymus serpyllum L. var. atlanticus Ball du Maroc. *Nat. Monsp. sér. Bot.* **16**, 161–175. 1965.

STAES, G. Revisie van het geslacht Thymus L. in België. *Bull. Jard. Bot. Bruxelles* **31**, 443–479. 1961.

WEBER, F. *Die Tschechoslowakischen Thymus-Arten und Opiz's Anteil an deren Erkennung*, in Klášterský, I., *P. M. Opiz und seine Bedeutung für die Pflanzentaxonomie*, 159–254. Praha. 1958.

HYSSOPUS

BENTHAM, G. *Labiatarum Genera et Species* **2**, 336–357. 1834.

SATUREJA

BENTHAM, G. *Labiatarum Genera et Species* **2**, 351–356. 1834: *loc. cit.* **3**, 729–730. 1835.

BRIQUET, J. Fragmenta Monographiae Labiatarum. *Ann. Conserv. Jard. Bot. Genéve* **2**, 186–192. 1896.

DE WOLF, G. P. Notes on cultivated Labiates, 4. Satureja and some related genera. *Baileya* **2**, 143–150. 1954.

EPLING, C. G. Synopsis of the genus Satureja. *Ann. Missouri Bot. Gard.* **14**, 47–86. 1927.

Epling, C. G. and Játiva, C. Revisión del género Satureja en America del sur. *Brittonia* **16**, 393–416. 1964.
Epling, C. G. and Játiva, C. A descriptive key to the species of Satureja indigenous to North America. *Brittonia* **18**, 244–248. 1966.

CALAMINTHA

Bentham, G. *Labiatarum Genera et Species* **2**, 386–388. 1834.
de Wolf, G. P. Notes on cultivated Labiates, 4. Satureja and some related genera. *Baileya* **2**, 143–150. 1954.

ACINOS

Bentham, G. *Labiatarum Genera et Species* **2**, 389–391. 1834.
de Wolf, G. P. Notes on cultivated Labiates, 4. Satureja and some related genera. *Baileya* **2**, 143–150. 1954.

CLINOPODIUM

Bentham, G. *Labiatarum Genera et Species* **2**, 391–393. 1834.

MELISSA

Bentham, G. *Labiatarum Genera et Species* **2**, 383–397. 1834: *loc. cit.* **3**, 732. 1835.

SALVIA

Bentham, G. *Labiatarum Genera et Species* **1**, 190–312. 1833: *loc. cit.* **3**, 717–726. 1835.
Briquet, J. Fragmenta Monographiae Labiatarum. *Bull. Herb. Boiss.* **2**, 135–138. 1894: *loc. cit.* **4**, 850–868. 1896: *Ann. Conserv. Bot. Gard. Genève* **2**, 123–176. 1896
Pénzes, A. Über die Ökologie und Systematik der Gruppe von Salvia verticillata L. *Borbasia* **5–6**, 1–31. 1946.

MELITTIS

Bentham, G. *Labiatarum Genera et Species* **2**, 503–504. 1834.
Klokov, M. Conspectus generis Melittis L. *Not. Syst.* (*Leningrad*) **18**, 183–217. 1957.

PRUNELLA

Bentham, G. *Labiatarum Genera et Species* **2**, 416–418. 1834.

STACHYS

Bentham, G. *Labiatarum Genera et Species* **2**, 525–566. 1834: *loc. cit.* **3**, 739–741. 1835.
Briquet, J. Fragmenta Monographiae Labiatarum. *Bull. Herb. Boiss.* **4**, 868–874. 1896: *Ann. Conserv. Jard. Bot. Genéve* **2**, 113–125. 1898.

BETONICA

BENTHAM, G. *Labiatarum Genera et Species* **2**, 532–534. 1834.

BALLOTA

BENTHAM, G. *Labiatarum Genera et Species* **3**, 592–600. 1834: *loc. cit.* 3, 743–744. 1835.

PATZAK, A. Revision der Gattung Ballota section Ballota. *Ann. Nat. Mus.* (*Wien*) **62**, 57–86. 1958

PATZAK, A. Revision der Gattung Ballota section Acanthoprasium und section Beringeria. *Ann. Nat. Mus.* (*Wien*) **63**, 33–81. 1959.

PATZAK, A. Zwei neue Ballota-Arten aus der Türkei nebst einem Nachtrag zur Gattung Ballota. *Ann. Nat. Mus.* (*Wien*) **64**, 42–56. 1961.

***LAMIASTRUM* (*GALEOBDOLON*)**

BENTHAM, G. *Labiatarum Genera et Species* **2**, 507–516. 1834.

LAMIUM

BENTHAM, G. *Labiatarum Genera et Species* **2**, 507–516. 1834: *loc. cit.* **3**, 738–739. 1835.

*DVOŘÁKOVÁ, M. Taxonomický přehled československých druhů rodu Lamium L. *Spisy Přirod. Univ. Purkyně Brně* **459**, 31–50. 1965.

LEONURUS

BENTHAM, G. *Labiatarum Genera et Species* **2**, 517–522. 1834: *loc. cit.* **3**, 739. 1835.

PHLOMIS

BENTHAM, G. *Labiatarum Genera et Species* **3**, 620–635. 1835.

GALEOPSIS

BENTHAM, G. *Labiatarum Genera et Species* **2**, 522–525. 1834.

BRIQUET, J. Monographie du genre Galeopsis. *Mém. cour. Mém. Sav. étr. Acad. Roy. Sci. Belg.* **1893**. Pp. xii + 323 + 51 figures. 1893.

BRIQUET, J. Fragmenta Monographiae Labiatorum. *Bull. Herb. Boiss.* **2**, 719–724. 1896.

HENRARD, J. Galeopsis, een systematisch-floristische studie. *Nederl. Kruidk. Arch.* **1919**, 158–188. 1919.

SLAVIKOVÁ, Z. Bemerkungen zur Taxonomie der Gattung Galeopsis L. *Nov. Bot. Hort. Bot. Univ. Prag.* **1963**, 39–43. 1963.

TOWNSEND, C. C. Some notes on Galeopsis ladanum L. and G. angustifolia Ehrh. ex Hoffm. *Watsonia* **5**, 143–149. 1962.

NEPETA

BENTHAM, G. *Labiatarum Genera et Species* **2**, 465–489. 1834: *loc. cit.* **3**, 734–738. 1835.

GLECHOMA

BENTHAM, G. *Labiatarum Genera et Species* **2**, 465–489. 1834.

KUPRIANOVA, L. The genus Glechoma L., and its species. *Bot. Žurn.* **33**, 230–238. 1948.

MARRUBIUM

BENTHAM, G. *Labiatarum Genera et Species* **3**, 587–592, 742–743. 1835.

SCUTELLARIA

BENTHAM, G. *Labiatarum Genera et Species* **2**, 419–445. 1834: *loc. cit.* **3**, 733. 1835.

EPLING, C. The American species of Scutellaria. *Univ. Calif. Publ. Bot.* **20**, 1–146. 1942.

*HAMILTON, A. *Esquisse d'une Monographie du genre Scutellaria, ou Toque . . . suive du Rétablissement du genre Scorodonia.* 1826–36.

TEUCRIUM

BÉGUINOT, A. Revisione monografica dei Teucrium della sezione Scorodonia (Adans.) Schreb. *Atti Accad. Sci. Veneto-Trentino-Istriana new sér.* **3**, 58–98. 1906.

BENTHAM, G. *Labiatarum Genera et Species* **3**, 660–692. 1835.

RECHINGER, K. H. Monographische Studie über Teucrium sect. Chamaedrys. *Bot. Archiv.* **42**, 335–420. 1941.

AJUGA

BENTHAM, G. *Labiatarum Genera et Species* **3**, 692–701. 1835.

PLANTAGINACEAE

BARNÉOUD, F. M. *Monographie générale de la Famille des Plantaginacées.* Pp. 52. Paris. 1845.

PLANTAGO

BARNÉOUD, F. M. *Monographie générale de la Famille des Plantaginacées.* Pp. 52. Paris. 1845.

BASSETT, I. J. The taxonomy of the North American Plantago L. section Micropsyllium Decne. *Canad. J. Bot.* **44**, 467–479. 1966.

CHALABI-KA'BI, Z. The genus Plantago in Iraq. *Bull. Iraq Mus.* **1** (2), 1–41. 1961.

PILGER, R. Über die Formen von Plantago major L. *Fedde. Rep.* **18**, 257–283. 1922.

PILGER, R. Beiträge zur Kenntnis der Gattung Plantago. *Fedde. Rep.* **18**, 449–475. 1922: *loc. cit.* **19**, 105–112, 114–119. 1923: *loc. cit.* **20**, 12–16. 1924: *loc. cit.* **21**, 97–102. 1925: *loc. cit.* **23**, 241–270. 1926: *loc. cit.* **34**, 147–166. 1933.

PILGER, R. Plantago coronopus und verwandte Arten. *Fedde. Rep.* **28**, 262–322. 1930.

PILGER, R. *Plantago*, in Engler, H. G. A. (Ed.), *Das Pflanzenreich* **102** (IV. 269), 39–432, 440–441. 1937.

LITTORELLA

BARNÉOUD, F. M. *Monographie générale de la Famille des Plantaginées.* Pp. 52. Paris. 1845.

FERNALD, M. L. North American Littorella. *Rhodora* **20**, 61–62. 1918.

PILGER, R. *Littorella*, in Engler, H. G. A. (Ed.), *Das Pflanzenreich* **102** (IV. 269), 433–437. 1937.

CAMPANULACEAE

BAILEY, L. H. and LAWRENCE, G. H. M. *The Garden of Bellflowers in North America.* Pp. xiii + 155 + 50 plates. New York. 1953.

WAHLENBERGIA

BAILEY, L. H. and LAWRENCE, G. H. M. *The Garden of Bellflowers in North America.* Pp. xiii + 155 + 50 plates. New York. 1953.

LOTHIAN, N. Critical notes on the genus Wahlenbergia Schrader: with descriptions of new species in the Australian region. *Proc. Linn. Soc. New South Wales* **71**, 201–236. 1947.

CAMPANULA

BAILEY, L. H. and LAWRENCE, G. H. M. *The Garden of Bellflowers in North America.* Pp. xiii + 155 + 50 plates. New York. 1953.

BEDDOME, R. H. An annotated list of the species of Campanula. *J. Roy. Hort. Soc.* **32**, 196–221. 1907.

BÖCHER, T. W. Experimental and cytological studies on plant species, 5. The Campanula rotundifolia complex. *Biol. Skr. Danske Vid. Selsk.* **11** (4), 1–69. 1960.

DAMBOLDT, J. Zytotaxonomische Revision der isophyllen Campanulae in Europa. *Bot. Jahrb.* **84**, 302–358. 1965.

DE CANDOLLE, A. *Monographie des Campanulées.* Pp. 384. Paris. 1830.

GADELLA, T. W. J. Cytotaxonomic studies in the genus Campanula. *Wentia* **11**, 1–104. 1964.

GUINOCHET, M. Recherches de taxonomie experimental sur la flore des Alpes et de la région mediterranéene occidentale, 2. Sur quelques formes du Campanula rotundifolia L.s.l. *Bull. Soc. Bot. France* **89**, 70–75, 153–156. 1942.

HRUBY, J. Campanulastudien. *Magyar Bot. Lap.* **29**, 152–276. 1930: *loc. cit.* **33**, 126–159. 1934: *Mitt. Fl.-Soz. Arbeitsgem.* **2**, 77–93. 1950.

PHITOS, D. Die quinquelokulären Campanula-Arten. *Österr. Bot. Zeitschr.* **112**, 449–498. 1965.

PODLECH, D. Beitrag zur Kenntnis der Subsektion Heterophylla (Witas.) Fed. der Gattung Campanula L. *Ber. Deutsch. Bot. Ges.* **75**, 237–244. 1962.

PODLECH, D. Revision der europäischen und nordafrikanischen Vertreter der Subsect. Heterophylla (Wit.) Fed. der Gattung Campanula. *Fedde. Rep.* **71**, 50–187. 1965.

SHETLER, S. G. A checklist and key to the species of Campanula native or commonly naturalised in North America. *Rhodora* **65**, 319–336. 1963.

WITASEK, J. Ein Beitrag zur Kenntnis der Gattung Campanula. *Abh. Zool.-Bot. Ges. Wien* **1** (3), 1–106. 1902.

LEGOUSIA

BAILEY, L. H. and LAWRENCE, G. H. M. *The Garden of Bellflowers in North America.* Pp. xiii + 155 + 50 plates. New York. 1953.

PHYTEUMA

SCHULZ, R. *Monographische Bearbeitung der Gattung Phyteuma.* Pp. 204. Geisenheim-am-Rhein. 1904.

LOBELIA

PRESL, C. B. Prodromus monographiae Lobeliacearum. *Abh. Böhm. Ges. Wiss.* (*Math.-Nat.*) *new ser.* **4**. Pp. 52. 1836.

WIMMER, F. E. Vorarbeiten zur Monographie der Campanulaceae-Lobelioideae, 2. Trib. Lobeliaceae. *Ann. Nat. Mus.* (*Wien*) **56**, 317–374. 1948.

WIMMER, F. E. *Lobelia,* in Engler, H. G. A. (Ed.), *Das Pflanzenreich* **106** (IV. 267b), 408–694. 1943: *loc. cit.* **107** (IV. 267b), 775–783. 1953.

RUBIACEAE

ASPERULA

ROMERIO, M. Contribution à la cytotaxonomie du groupe de l'Asperula cynanchica L. *Bull. Soc. Neuchât. Sci. Nat.* **88**, 65–76. 1965.

GALIUM

EHRENDORFER, F. Neufassung der Sektion Lepto-Galium Lange und Beschreibung neuer Arten und Kombinationen. *Sitz. Österr. Akad. Wiss. Math.-Nat. Kl. Abt.* **1** (169), 407–421. 1961.

GABRIËLS, J. Revisie der geslachten Galium L. en Cruciata Mill. (Rubiaceae) in België. *Bull. Jard. Bot. Bruxelles* **35**, 109–166. 1965.

HAYEK, A. von. *Galium,* in Hegi, G. *Ill. Fl. Mitteleur.* **6** (1), 207–231. 1914.

LÖVE, A. and LÖVE, D. Cytotaxonomic studies on the northern Bedstraw. *Amer. Midl. Nat.* **52**, 88–105. 1954.

UBACH, M. Estudio anatómico de la epidermis del fruto algunas especies de Galium. *Collect. Bot.* **3**, 110–135. 1951.

CRUCIATA

GABRIËLS, J. Revisie der geslachten Galium L. en Cruciata Mill. (Rubiaceae) in België. *Bull. Jard. Bot. Bruxelles* **35**, 109–166. 1965.

CRUCIANELLA

MALINOWSKI, E. Les espèces du genre Crucianella L. *Bull. Soc. Bot. Genève ser.* 2 **1910**, 9–17. 1910.

CAPRIFOLIACEAE

SAMBUCUS

SCHWERIN, F. von. Monographie der Gattung Sambucus. *Mitt. Deutsch. Dendr. Ges.* **1909**. Pp. 56. 1909.

SCHWERIN, F. von. Revisio generis Sambucus. *Mitt. Deutsch. Dendr. Ges.* **1920**, 194–231. 1920.

VIBURNUM

MCATEE, W. L. *A Review of the Neoarctic Viburnum.* Pp. iv + 9 plates. Chapel Hill. 1956.

SYMPHORICARPOS

GRAY, A. Revision of the genus Symphoricarpos. *J. Linn. Soc. Bot.* **14**, 9–12. 1875.

JONES, G. N. A monograph of the genus Symphoricarpos. *J. Arnold Arb.* **21**, 201–252. 1940.

LINNAEA

GIGER, C. Linnaea borealis L. eine monographische Studie. *Beih. Bot. Centr.* **30** (2), 1–78. 1913.

WITTROCK, V. B. Linnaea borealis L. Species polymorpha et polychroma. *Acta Hort. Berg.* **4** (7). Pp. 187. 1907.

LONICERA

REHDER, A. Synopsis of the genus Lonicera. *Ann. Rep. Missouri Bot. Gard.* **14**, 27–232. 1903.

ADOXACEAE

ADOXA

*STURM, K. Monographische Studien über Adoxa moschatellina L. *Viert. Naturf. Ges. Zürich* **55**, 391–462. 1911

VALERIANACEAE

VALERIANELLA

DYAL, S. G. Valerianella in North America. *Rhodora* **40**, 185–212. 1938.

KROK, T. O. B. N. Anteckningar till en Monografi öfver Vaxtfamiljen Valerianeae . . . 1. Valerianella Hall. *Kung. Svensk. Vet.-Akad. Hand.* **5**, 1–105. 1864.

WOODS, J. Observations of the species of Fedia. *Trans. Linn. Soc.* **17**, 421–433. 1836.

VALERIANA

ČERVENKA, V. J. B. Study of polyploid forms of Valeriana officinalis. *Preslia* **27**, 234–242. 1955.

DRABBLE, E. Valeriana officinalis and its allies in Great Britain. *Rep. Bot. Soc. & E.C.* **10**, 249–257. 1933.

HEGNAUER, R. and MEIJERS, T. Valeriana officinalis in Holland. *Planta Medica* **6**, 349–372. 1958.

LAWALRÉE, A. Le groupe du Valeriana officinalis L. en Belgique. *Bull. Jard. Bot. Bruxelles* **22**, 193–200. 1952.

MEIJERS, T. *Een Onderzoek van het Linnean Valeriana officinalis L. in Nederland.* Pp. 112. S'Gravenhage. 1957.

SKALIŃSKA, M. Studies in cytoecology, geographic distribution and evolution of Valeriana L. *Bull. Acad. Polon. Sci. Lett. ser. B* 1 **1950**, 149–175. 1951.

SPRAGUE, T. A. The British forms of Valeriana officinalis. *Watsonia* **2**, 145–147. 1952.

VOROSHILOV, V. N. *Valeriana officinalis*. Pp. 159. Academy of Sciences URSS. Moscow. 1959.

WALTHER, E. Zur Morphologie und Systematik des Arzneibaldrians in Mitteleuropa. *Mitt. Thür. Bot. Ges. Beih.* **1**, 1–108. 1949.

WALTHER, E. Valeriana-Studien, 1. Valeriana montana L. und Valeriana tripteris L. *Mitt. Thür. Bot. Ges.* **1**, 144–165. 1949.

DIPSACACEAE

COULTER, T. Mémoire sur les Dipsacacées. *Mém. Soc. Phys. Genève* **2**, 13–60 + tab. 2. 1824.

TORTAJADA, J. M. Contribución al estudio de las Dipsacaceas Españolas. *Bol. Inst. Nac. Invest. Agron.* **24**, 123–216. 1964.

KNAUTIA

*BORBÁS, V. *Revisio Knautiarum.* Pp. 90. Kolozsvár. 1904.

COULTER, T. Mémoire sur les Dipsacacées. *Mém. Soc. Phys. Genève* **2**, 13–60 + tab. 2. 1824.

SZABÓ, Z. Monographie der Gattung Knautia. *Bot. Jahrb.* **36**, 389–442. 1905.

SZABÓ, Z. A Knautia génusz monográphiája. *Math. Term. Közl.* **31**, 1–436. 1911.

TORTAJADA, J. M. Contribución al estudio de las Dipsacaceas Españolas. *Bol. Inst. Nac. Invest. Agron.* **24**, 123–216. 1964.

SCABIOSA

Coulter, T. Mémoire sur les Dipsacacées. *Mém. Soc. Phys. Genève* **2**, 13–60 + tab. 2.1824.

Tortajada, J. M. Contribución al estudio de las Dipsacaceas Españolas. *Bol. Inst. Nac. Invest. Agron.* **24**, 123–216.1964.

SUCCISA

Baksay, L. Monographie der Gattung Succisa. *Ann. Mus. Hung. new ser.* **2**, 237–259.1952.

Coulter, T. Mémoire sur les Dipsacacées. *Mém. Soc. Phys. Genève* **2**, 13–60 + tab. 2.1824.

Tortajada, J. M. Contribución al estudio de las Dipsacaceas Españolas. *Bol. Inst. Nac. Invest. Agron.* **24**, 123–216.1964.

CEPHALARIA

Coulter, T. Mémoire sur les Dipsacacées. *Mém. Soc. Phys. Genève* **2**, 13–60 + tab. 2.1824.

Szabó, Z. A Cephalaria-Génusz monográfiája. *Math. Term. Közl.* **38** (4). Pp. viii + 352.1940.

Tortajada, J. M. Contribución al estudio de las Dipsacaceas Españolas. *Bol. Inst. Nac. Invest. Agron.* **24**, 123–216.1964.

COMPOSITAE

Lessing, C. F. *Synopsis Generum Compositarum earumque Dispositionis Novae tentamen Monographiis Multarum Capensium Interjectis.* Pp. xi + 473. Berolini. 1832.

CALENDULA

Lanza, D. Monografio del genero Calendula L. *Atti Reale Accad.* (*Palermo*) *ser.* 3 **12**, 1–166.1923.

Meusel, H. and Ohle, H. Zur Taxonomie und Cytologie der Gattung Calendula. *Österr. Bot. Zeitschr.* **113**, 191–210.1966.

BIDENS

Sherff, E. E. The genus Bidens. *Publ. Field Mus. Nat. Hist. Bot. Ser.* **16**, 6–346: *loc. cit.* **17**, 347–709.1937.

SIGESBECKIA

Henker, H. Die Gattung Sigesbeckia L. in Europe unter besonderer Berücksichtigung von Deutschland. *Arch. Freunde Nat. Mecklenburg* **11**, 7–54.1965.

GALINSOGA

ST. JOHN, H. and WHITE, D. The genus Galinsoga in North America. *Rhodora* **22**, 97–101. 1920.

THELLUNG, A. Über die in Mitteleuropa vorkommenden Galinsoga-Formen. *Allgem. Bot. Zeitschr.* **21**, 1–16. 1915.

AMBROSIA

LAWALRÉE, A. Les Ambrosia adventices en Europe occidentale. *Bull. Jard. Bot. Bruxelles* **18**, 305–315. 1947.

LAWALRÉE, A. Note complémentaire sur les Ambrosia adventices en Europe occidentale. *Bull. Soc. Bot. Belg.* **87**, 207–208. 1955.

PAYNE, W. W. A re-evaluation of the genus Ambrosia (Compositae). *J. Arnold Arb.* **45**, 401–438. 1964.

XANTHIUM

HAYEK, A. von. *Xanthium*, in Hegi, G., *Illustr. Fl. Mitteleur.* **6** (1), 498–503. 1918.

LÖVE, D. and DANSEREAU, P. Biosystematic studies on Xanthium: taxonomic appraisal and ecological status. *Canad. J. Bot.* **37**, 173–208. 1959.

MILLSPAUGH, C. F. and SHERFF, E. E. Revision of the North American species of Xanthium. *Publ. Field Mus. Nat. Hist. Bot. Ser.* **4** (2), 9–49 + 18 plates. 1919.

WALLROTH, K. F. W. Monographischer Versuch über die Gewächs-Gattung Xanthium Diosc. *Beitr. Bot. (Leipzig)* **1**, 219–244. 1844.

WEIN, K. Beiträge zur Geschichte der Einfuhrung und Einbürgerung einiger Arten von Xanthium in Europa. *Beih. Bot. Centr.* **42**(2), 151–176. 1925.

WIDDER, F. J. Die Arten der Gattung Xanthium. *Fedde. Rep. Beih.* **20**, 1–223. 1923.

WIDDER, F. J. Übersicht über die bisher in Europa beobachteten Xanthium —Arten und Bastarde. *Fedde. Rep.* **21**, 273–305. 1925.

HELIANTHUS

WATSON, E. E. Contributions to a monograph of the genus Helianthus. *Papers Michigan Acad.* **9**, 305–475. 1929.

VERBESINA

COLEMAN, J. R. A taxonomic revision of Section Ximenesia of the genus Verbesina L. (Compositae). *Amer. Midl. Nat.* **76**, 475–481. 1966.

ROBINSON, B. L. and GREENMAN, J. H. A synopsis of the genus Verbesina with an analytical key to the species. *Proc. Amer. Acad. Arts Sci.* **34**, 534–564. 1899.

SCHKUHRIA

REISER, C. B. Jnr. A revision of the genus Schkuhria. *Ann. Missouri. Bot. Gard.* **32**, 265–278. 1945.

COSMOS

SHERFF, E. E. Revision of the genus Cosmos. *Publ. Field Mus. Nat. Hist. Bot. Ser.* **7** (6), 401–447. 1932.

SENECIO

CUFODONTIS, G. Kritische Revision von Senecio sect. Tephroseris. *Fedde. Rep. Beih.* **70**, 1–226. 1933.

GREENMAN, J. H. Monographie der nord- und central amerikanischen Arten der Gattung Senecio. *Bot. Jahrb.* **32**, 1–33. 1903.

GREENMAN, J. H. Monograph of the North and Central American species of the genus Senecio, 2. *Ann. Missouri Bot. Gard.* **2**, 573–626. 1915: *loc. cit.* **3**, 85–194. 1916: *loc. cit.* **4**, 15–35. 1917: *loc. cit.* **5**, 37–107. 1918.

MOSSERAY, R. Materiaux pour une Flora de Belgique, 5. Genre Senecio. *Bull. Jard. Bot. Bruxelles* **14**, 57–82. 1936.

DORONICUM

CAVILLIER, F. Étude sur les Doronicum à fruits homomorphes. *Ann. Conserv. Gard. Bot. Genève* **10**, 177–251. 1907.

CAVILLIER, F. Nouvelles études sur le genre Doronicum. *Ann. Conserv. Gard. Bot. Genève* **13–14**, 195–368. 1911.

ROUY, M. G. Le genre Doronicum dans la flora européenne et dans la flora atlantique. *Rev. Bot. Syst. Géogr. Bot.* **1**, 17–22, 33–40, 49–56. 1903.

INULA

BECK, G. Inulae Europae. Die europäischen Inula-Arten. *Denkschr. Akad. Wiss. Math.-Nat. Kl. (Wien)* **44**, 283–339. 1881.

BORBÁS, V. Die ungarischen Inula-Arten, besonders aus der Gruppe Enula. *Bot. Jahrb.* **8**, 222–243. 1887.

ANTENNARIA

BORISSOVA, A. T. De genere Antennaria Gaertn. notulae systematicae. *Not. Syst. (Leningrad)* **20**, 289–295. 1960.

CHRTEK, J. and POUZAR, Z. Observations on some Scandinavian species of the Antennaria Gaertn. genus. *Nov. Bot. Hort. Bot. Univ. Prag.* **1961**, 11–15. 1961.

CHRTEK, J. and POUZAR, Z. A contribution to the taxonomy of some European species of the genus Antennaria Gaertn. *Acta Univ. Carolinae* **1962**, 105–136. 1962.

PORSILD, A. E. The genus Antennaria in north western Canada. *Canad. Field-Nat.* **64**, 1–25. 1950.

PORSILD, A. E. The genus Antennaria in eastern Arctic and subarctic America. *Bot. Tidsskr.* **61**, 22–56. 1965.

FILAGO

WAGENITZ, G. Zur Systematik und Nomenklatur einiger Arten von Filago L. emend. Gaertn. subgen. Filago ("Filago germanica" Gruppe). *Willdenowia* **4**, 37–59. 1965.

SOLIDAGO

BEAUDRY, J. R. and CHABOT, D. L. Studies on Solidago L., 1. S. altissima L. and S. canadensis L. *Contr. Inst. Bot. Univ. Montréal* **70**, 65–72. 1957.

ROUY, G. Le Solidago virga-aurea L. dans la flore française. *Rev. Bot. Syst. Géogr. Bot.* **1**, 1–13. 1903.

ASTER

BRITTON, C. E. The naturalised and alien Asters of the British Plant List, Ed. 2. *Rep. Bot. Soc. & E.C.* **9**, 710–718. 1932.

BURGESS, E. Species and variations of Biotian Asters, with discussion of variability in Aster. *Mem. Torrey Bot. Club* **13**, 1–149. 1906.

CRONQUIST, A. Revision of the western North American species of Aster centering about Aster foliaceous Lind. *Amer. Midl. Nat.* **29**, 429–468. 1943.

NEES VON ESENBECK, C. G. *Synopsis specierum generis Aster herbacearum, praemiss nonnullis de Asteribus in genere*. Pp. 32. Erlangae. 1818.

NEES VON ESENBECK, C. G. *Genera et Species Asterearum*. Pp. xiv + 300. Vratislaviae. 1832.

ONNO, M. Geographisch-morphologische Studien über Aster alpinus L. und verwandte Arten. *Bibl. Bot.* **106**, 1–83. 1932.

THELLUNG, A. Die in Mitteleuropa kultivierten und verwildearten Aster-und-Helianthus-Arten. *Allgem. Bot. Zeitschr.* **19**, 87–89, 101–112, 132–140. 1913.

ERIGERON

CRONQUIST, A. Revision of the North American species of Erigeron north of Mexico. *Brittonia* **6**, 121–302. 1947.

SOLBRIG, O. T. The South American species of Erigeron. *Contr. Gray Herb.* **191**, 3–79. 1963.

VIERHAPPER, F. Monographie der alpinen Erigeron-Arten europas und Vorder-asiens. *Beih. Bot. Centr.* **19** (2), 385–560. 1906.

CONYZA

CRONQUIST, A. The separation of Erigeron from Conyza. *Bull. Torrey Bot. Club* **70**, 629–632. 1943.

ACHILLEA

HEIMER, A. Monographia sectionis "Ptarmica" Achillea generis. Die Arten XXX der Section Ptarmica. *Denkschr. Akad. Wiss. Math.-Nat. Kl. (Wien)* **48**, 113–192. 1894.

MULLIGAN, G. A. and BASSETT, I. J. Achillea millefolium complex in Canada and adjacent proportions of the United States. *Canad. J. Bot.* **37**, 73–79. 1959.

PRODAN, J. Achilleae Romaniae et descriptio aliquet specierum e peninsula balcanica nostris speciebus propinquarum facta. *Bul. Acad. Inalt. Stud. Agron. Cluj Memorii* **2**. Pp. iv + 68 + 43 plates + 3 maps. 1931.

*SPUDILOVA, V. A monographical study of the genus Achillea in Czechoslovakia. *Acta Rer. Nat. Distr. Ostrav.* **18**, 101–106, 190–199. 1957.

CHRYSANTHEMUM

BÖCHER, T. W. and LARSEN, K. Cytotaxonomical studies in the Chrysanthemum leucanthemum complex. *Watsonia* **4**, 11–16. 1957.

FAVARGER, C. Distribution en Suisse des races chromosomiques de Chrysanthemum L. *Ber. Schweiz. Bot. Ges.* **69**, 26–46. 1959.

HORVATIĆ, S. Neuer Beitrag zur Kenntnis der Leucanthemum Formen in der Flora Jugoslaviens. *Acta Bot. Zagreb.* **10**, 61–100. 1935.

MULLIGAN, G. A. Chromosome races in the Chrysanthemum leucanthemum complex. *Rhodora* **60**, 122–125. 1958.

COTULA

EDGAR, E. Studies in New Zealand Cotulas. *Trans. Roy. Soc. N.Z.* **85**, 375–377. 1958.

CALOTIS

DAVIS, G. L. Revision of the genus Calotis R. Br. *Proc. Linn. Soc. New South Wales* **77**, 146–188. 1952.

ARTEMISIA

BESSER, W. S. Synopsis des Absinthes. *Bull. Soc. Nat. Moscou* **1** (8), 219–266. 1829.

BESSER, W. S. Tentamen de Abotanis seu de Sectione 2. Artemisiarum Linnaei. *Nouv. Mém. Soc. Nat. Moscou* **3**, 5–92. 1834.

BESSER, W. S. De Seriphidiis seu de Sectione 3. Artemisiarum Linn. *Bull. Soc. Nat. Moscou* **7**. Pp. 46. 1834.

BESSER, W. S. Dracunculi seu de Sectione 4 et ultima Artemisiarum Linnaei. *Bull. Soc. Nat. Moscou* **8**. Pp. 95. 1835.

BESSER, W. S. Supplementum ad synopsis Absynthiorum tentamen de Abrotanis dissertationem de Seriphidiis atque de Dracunculus. *Bull. Soc. Nat. Moscou* **9**. Pp. 115. 1836.

BESSER, W. S. Revisio Artemisiarum Museii Regii Berolinensis, Cuius partem constituit herbarium Willdenovianum Instituta. *Linnaea* **15**, 83–111. 1841.

BESSER, W. S. Monographiae Artemisiarum Sectio 1. Dracunculi Frutescentes. *Mém. Acad. Sci. St. Pétersb. (Sci. Phys. Math.)* **1845**, 1–44. 1845.

Keck, D. A revision of the Artemisia vulgaris complex in North America *Proc. California Acad. Sci. ser.* 4 **25**, 421–468.1946.

Wendelberger, G. Die Sektion Heterophyllae der Gattung Artemisia. *Bibl. Bot.* **125**. Pp. v + 193.1960.

ECHINOPS

Bornmüller, J. Revisions-Ergebnisse einiger orientalischer und zentralasiatischer Arten der Gattung Echinops. *Beih. Bot. Centr.* **36** (2), 200–228.1918.

Bunge, A. Über die Gattung Echinops. *Bull. Acad. Sci. St. Pétersb.* **6**, 390–412.1863.

IVA

Jackson, R. C. A revision of the genus Iva L. *Kansas Univ. Sci. Bull.* **41**, 793–876.1960.

ARCTIUM

Arènes, J. Monographie du genre Arctium L. *Bull. Jard. Bot. Bruxelles* **20**, 67–156.1950.

Babington, C. C. On the British species of Arctium. *Ann. Mag. Nat. Hist. ser.* 2 **17**, 369–377.1856.

de Langhe, J. E. Les Bardanes (Genre Arctium) de Belgique et des régions voisines. *Nat. Belg.* **47**, 21–30.1965.

Evans, A. H. The British species of Arctium. *J. Bot.* **51**, 113–119.1913.

*Wilpert, H. *Vorarbeiten zu einer Monographie der Gattung Arctium.* Oppela. 1928.

CARDUUS

Kazmi, S. M. A. Revision der Gattung Carduus (Compositae). *Mitt. Bot. Staats. München* **5**, 139–198.1963: *loc. cit.* **5**, 279–550.1964.

CIRSIUM

Gross-Würzburg, L. and Gugler, W. Ueber unter fränkische Cirsien. *Allgem. Bot. Zeitschr.* **10**, 66–70, 112–119, 129–135.1904.

Naegeli, K. *Die Cirsien der Schweiz.* Pp. 166. Neuchâtel. 1841.

Petrak, F. Der Formenkreis des Cirsium eriophorum (L.) Scop. in Europa. *Bibl. Bot.* **78**. Pp. 92 + Taf. 6.1912.

Petrak, F. Die nordamerikanische Arten der Gattung Cirsium. *Beih. Bot. Centr.* **35** (1), 223–567.1917.

Petrak, F. Über einige Arten und Bastarde der Gattung Cirsium. *Mitt. Thür. Bot. Ges.* **2**, 13–41.1960.

Rouy, M. G. Conspectus des espèces, sous espèces, formes, varietés et hybrides du genre Cirsium dans la flore française. *Rev. Bot. Syst. Géogr. Bot.* **2**,1–11, 28–32, 42–47,57–62, 74–78, 115–118.1904.

ONOPORDUM

DRESS, W. J. Notes on the cultivated Compositae, 9. Onopordum. *Baileya* **14**, 75–86. 1966.

ROUY, G. Revision du genre Onopordon. *Bull. Soc. Bot. France* **43**, 577–599. 1896.

CENTAUREA

ARÉNES, J. Les Centaurées de la sous-section Jacea: systématique, chorologie et phylogénie. *Bull. Jard. Bot. Bruxelles* **27**, 143–157. 1957.

BRIQUET, J. *Monographie des Centaurées des Alpes maritimes.* Pp. 196. Basel. 1902.

BRITTON, C. E. British Centaureas of the nigra group. *Rep. Bot. Soc. & E.C.* **6**, 406–417. 1922.

GUGLER, W. Die Centaureen des ungarischen National-museums. Vorarbeiten zu einer Monographie der Gattung Centaurea und ihr nächst verwandten Genera. *Ann. Mus. Hung.* **6**, 15–297 + Tab. 1. 1908.

HAYEK, A. von. Die Centaurea-Arten Österreich-Ungarns. *Denkschr. Akad. Wiss. Math.-Nat. Kl. (Wien)* **72**, 585–773. 1901.

MARSDEN-JONES, E. M. and TURRILL, W. B. *British Knapweeds. A Study in Synthetic Taxonomy.* Pp. xiii + 201. Ray Society. London. 1954.

PRODAN, I. Specii din Genul Centaurea aflate și Studiate in Herbarul Universitătii din Cluj. *An. Acad. Rep. Pop. Române* **3**, 1–34. 1950.

ROUY, M. G. Les Centaurea de la Section Acrolophus dans la flora française. *Rev. Bot. Syst. Géogr. Bot.* **2**, 140–149, 156–163. 1904.

CARTHAMUS

HANELT, M. P. Monographische Übersicht der Gattung Carthamus L. (Compositae). *Fedde. Rep.* **67**, 41–180. 1962.

LAPSANA

WALLROTH, K. F. W. Monographischer Versuch über die Gewachs-Gattung Lampsana Dodon. *Beitr. Bot. (Leipzig)*, 127–139. 1842.

HYPOCHOERIS

SCHULTZ, C. H. B. Hypochoerideae. *Nova Acta Acad. Leop.-Carol.* **21**, 85–172. 1845.

LEONTODON

BALL, J. Outlines of a monograph of the genus Leontodon. *Ann. Mag. Nat. Hist. ser.* 2 **6**, 1–18. 1850.

CSONGOR, G. Monographie critique des espèces du genre Leontodon dans le Bassins Carpatiques. *Acta Geobot. Hung. ser. nov.* **1**, 51–69. 1947.

TRAGOPOGON

REGEL, C. von. Die Verbreitung einiger europäisch-asiatischer Tragopon-Arten. *Ber. Schweiz. Bot. Ges.* **65**, 251–262. 1955.

SCORZONERA

*LIPSCHITZ, S. J. *Fragmenta Monographiae Generis Scorzonera*. Pp. 164. Moscow. 1935: pars. 2. Pp. 167. 1939.

TIMBAL-LAGRAVE, M. E. *Essai Monographique sur les espéces du genre Scorzonera L. de la flore française*. Pp. 16. Toulouse. 1886.

LACTUCA

LINDQVIST, K. On the origin of the cultivated lettuce. *Hereditas* **46**, 319–350. 1960.

SONCHUS

BOULOS, L. Cytotaxonomic studies in the genus Sonchus, 2. The genus Sonchus, a general systematic treatment. *Bot. Not.* **113**, 400–420. 1960.

CICERBITA

BEAUVERD, G. Le genre Cicerbita. *Bull. Soc. Bot. Genève sér.* **2** 1910, 99–144. 1910.

HIERACIUM

ARVET-TOUVET, C. *Hieraciorum praesertion Galliae et Hispaniae Catalogus Systematicae*. Pp. ix + 480. Paris. 1913.

BACKHOUSE, J., Jnr. *A Monograph of the British Hieracia*. Pp. viii + 9–89. York. 1856.

DAHLSTEDT, H. De Hieraciis nonnullis Scandinavicis in Horto Bergiano cultis. *Acta Hort. Berg.* **1** (7), 3–146. 1891.

DAHLSTEDT, H. *Hieracium*, in Lindman, C. A. M., *Svensk Fanerogamflora*, 589–628. Stockholm. 1918.

*FRIES, E. M. Symbolae ad Historiam Hieraciorum. *Nova Acta Regiae Soc. Sci. Upsal.* **13–14**. 1848.

*FRIES, E. M. *Epicrisis Generis Hieraciorum*. Pp. 158. Upsala. 1862.

HANBURY, F. J. *An Illustrated Monograph of the British Hieracia*. Pp. 68. London. 1889–98.

JOHANSSON, K. and SAMUELSSON, G. *Dalarnes Hieracia Silvaticiforma*. Pp. 93. Leipzig. 1922.

JOHANSSON, K. and SAMUELSSON, G. *Dalarnes Hieracia Vulgatiforma*. Pp. 81. Leipzig. 1923.

LINTON, W. R. *An Account of the British Hieracia*. Pp. viii + 96. London. 1905.

PUGSLEY, H. W. A Prodromus of the British Hieracia. *J. Linn. Soc. Bot.* **54**, 1–356. 1948.

SELL, P. D. and WEST, C. Notes on British Hieracia, 1. *Watsonia* **3**, 233–236. 1955: *loc. cit.* **5**, 215–223. 1962.

SELL, P. D. and WEST, C. A revision of the British species of Hieracium section Alpestria (Fries) F. N. Williams. *Watsonia* **6**, 85–105. 1965.

ZAHN, K. Die Hieracien der Schweiz. *Ber. Schweiz. Bot. Ges.* **1906**, 163–728. 1906.

*ZAHN, K. *Hieraciotheca Europaea: Schedae ad centurian.* Karlsruhe. 1912–14.

ZAHN, K. *Les Hieracium des Alpes Maritimes.* Pp. vii + 404. Lyon. 1916.

ZAHN, K. *Hieracium*, in Engler, H. G. A. (Ed.), *Das Pflanzenreich* **75** (IV. 280), 1–288: *loc. cit.* **76** (IV. 280), 289–576: *loc. cit.* **77** (IV. 280), 577–864. 1921: *loc. cit.* **79** (IV. 280), 865–1146. 1922: *loc. cit.* **82** (IV. 280), 1147–1705. 1923.

PILOSELLA (see also under ***HIERACIUM***)

DIJKSTRA, S. J., KERN, J. H., REICHGELT, T. and VAN SOEST, J. L. Sur quelques Hieracia subg. Pilosella des Pays-Bas. *Acta Bot. Neerl.* **2**, 522–534. 1953.

NAEGELI, C. von and PETER, A. *Die Hieracien Mittel-Europas: Monographische Bearbeitung der Piloselloiden.* Pp. xi + 931. München. 1885. Vol. 2. *Monographische Bearbeitung der Archieracien.* Pp. 240. 1886: Pp. 241–340. 1889.

OMANG, S. O. F. *Die Hieracien Norwegens*, 1. *Monographische Bearbeitung der untergattung Piloselloidea.* Pp. 179. Oslo. 1935.

PETER, A. Über spontane und künstliche Gartenbastarde der Gattung Hieracium sect. Piloselloidea. *Bot. Jahrb.* **6**, 111–136. 1885.

CREPIS

BABCOCK, E. The genus Crepis. *Univ. Calif. Publ. Bot.* **21–22**, 1–1030. 1947.

DRABBLE, E. The genus Crepis in Great Britain. *J. Bot.* **70**, 274–281. 1932: *loc. cit.* **71**, 57–63. 1933.

TARAXACUM

CHRISTIANSEN, M. P. Nye Taraxacum-Arter af Gruppen Vulgaria. *Dansk Bot. Arkiv* **9** (2). Pp. 31 + Taf. 23. 1936.

DAHLSTEDT, H. Om skandinaviska Taraxacum former. *Bot. Not.* **1905**, 145–172. 1905.

DAHLSTEDT, H. Studier öfver arktiska Taraxaca. *Arkiv Bot.* **4** (8). Pp. 41. 1905.

DAHLSTEDT, H. Arktiska och alpina arter inom formgruppen Taraxacum ceratophorum (Led.) DC. *Arkiv Bot.* **5** (9). Pp. 44 + Taf. 18. 1906.

DAHLSTEDT, H. *Einige Wildwachende Taraxaca aus dem Botanischen Garten zu Upsala*, in *Botaniska Studier tillägnade F. R. Kjellman den 4 November* 1906, 164–183. Uppsala. 1906.

DAHLSTEDT, H. Taraxacum palustre (Ehrh.) und Verwandte Arten in Skandinavien. *Arkiv Bot.* **7** (6). Pp. 29 + Taf. 4. 1907.

DAHLSTEDT, H. Ueber einige in Bergianschen Botanischen Garten in Stockholm kultivierte Taraxaca. *Acta Hort. Berg.* **4** (2), 3–31 + 2 plates. 1907.

DAHLSTEDT, H. Nya skandinaviska Taraxacum-arter, jämte öfversikt af grupperna Erythrosperma och Obliqua. *Bot. Not.* **1909**, 167–179. 1909.

DAHLSTEDT, H. *Östsvenska Taraxaca.* Pp. 74. Uppsala. 1910.

DAHLSTEDT, H. Nya östsvenska Taraxaca. *Arkiv Bot.* **10** (6). Pp. 36. 1911.

DAHLSTEDT, H. Västsvenska Taraxaca. *Arkiv Bot.* **10** (11). Pp. 74. 1911.

DAHLSTEDT, H. Nordsvenska Taraxaca. *Arkiv Bot.* **12** (2). Pp. 122. 1912.

DAHLSTEDT, H. *Taraxacum*, in Lindman, C. A. M. *Svensk Fanerogamflora*, 559–589. Stockholm. 1918.

DAHLSTEDT, H. De Svenska arterna av släktet Taraxacum, 1. Erythrosperma. 2. Obliqua. *Acta Flora Suec.* **1**, 1–160 + Taf. 11. 1921.

DAHLSTEDT, H. De Svenska arterna av släktet Taraxacum, 3. Dissimilia. 4. Palustria. 5. Ceratophora. 6. Arctica. 7. Glabra. *Kung. Svenska Vet.-Akad. Handl. Tredje serien* **6** (3). Pp. 66 + 16 plates. 1928.

HANDEL-MAZZETTI, H. F. von. *Monographie der Gattung Taraxacum.* Pp. 175. Leipzig and Wien. 1908.

HANDEL-MAZZETTI, H. F. von. Nachträge zur Monographie der Gattung Taraxacum. *Österr. Bot. Zeitschr.* **72**, 254–275. 1923.

VAN SOEST, J. L. Het geslacht Taraxacum in Nederland, 1. (Obliqua, Dissimilia, Erythrospermum). *Nederl. Kruidk. Arch.* **49**, 213–237. 1939: 2. Palutria, Spectabilia, enkele Vulgaris. *loc. cit.* **52**, 215–236. 1942.

VAN SOEST, J. L. Sur quelques Taraxacum et Hieracia du Portugal. *Agron. Lusit.* **10**, 6–23. 1948.

VAN SOEST, J. L. Sur quelques Taraxacum du Portugal. *Agron. Lusit.* **13**, 67–76. 1951.

VAN SOEST, J. L. Taraxacum section Vulgaria Dt. in Nederland. *Acta Bot. Neerl.* **4**, 82–107. 1955.

VAN SOEST, J. L. Les Taraxacum de Belgique. *Bull. Jard. Bot. Bruxelles* **26**, 211–235. 1956: *loc. cit.* **31**, 319–389. 1961.

VAN SOEST, J. L. Taraxacum sectio Obliqua Dt. en sectio Erythrosperma Dt. em Lb. in Nederland. *Acta Bot. Neerl.* **6**, 74–92. 1957.

VAN SOEST, J. L. Alpine species of Taraxacum with special reference to the central and eastern Alps. *Acta Bot. Neerl.* **8**, 77–138. 1959.

VAN SOEST, J. L. Quelques nouvelles espèces de Taraxacum, natives d'Europe. *Acta Bot. Neerl.* **10**, 280–306. 1961.

VAN SOEST, J. L. Taraxacum sect. Palustria Dahlstedt. *Acta Bot. Neerl.* **14**, 1–53. 1965.

VAN SOEST, J. L. Korte determinatietabel voor de Nederlandse soorten van Taraxacum sect. Erythrosperma Dahlst. em. Lindb. f. *Gorteria* **3**, 43–47. 1966.

MONOCOTYLEDONS

ALISMATACEAE

BEAL, O. E. The Alismataceae of the Carolinas. *J. Elisha Mitchell Sci. Soc.* **76**, 68–79. 1960.

BUCHENAU, F. Index criticus Butomacearum, Alismacearum, Juncaginacearumque hucusque descriptarum. *Abh. Nat. Ver. Bremen* **1**. Pp. 61. 1868.

GLÜCK, H. *Biologische und morphologische Untersuchungen über Wasser-und Sumpfgewächse*, 1. *Die Lebensgeschichte der europäischen Alismaceen*. Pp. xxiv + 312 + Taf. 7. Jena. 1905.

BALDELLIA

BEAL, O. E. The Alismataceae of the Carolinas. *J. Elisha Mitchell Sci. Soc.* **76**, 68–79. 1960.

BUCHENAU, F. Index criticus Butomacearum, Alismacearum, Juncaginacearumque hucusque descriptarum. *Abh. Nat. Ver. Bremen* **1**. Pp. 61. 1868.

BUCHENAU, F. *Baldellia*, in Engler, H. G. A. (Ed.), *Das Pflanzenreich* **16** (IV. 15), 25–35. 1903.

GLÜCK, H. *Biologische und morphologische Untersuchungen über Wasser- und Sumpfgewächse*, 1. *Die Lebensgeschichte der europäischen Alismaceen*. Pp. xxiv + 312 + Taf. 7. Jena. 1905.

MICHELI, M. *Echinodorus*, in de Candolle, A. and de Candolle, C., *Monographie Phanerogamarum* **3**, 44–60. Paris. 1881.

LURONIUM

BEAL, O. E. The Alismataceae of the Carolinas. *J. Elisha Mitchell Sci. Soc.* **76**, 68–79. 1960.

BUCHENAU, F. Index criticus Butomacearum, Alismacearum, Juncaginacearumque hucusque descriptarum. *Abh. Nat. Ver. Bremen* **1**. Pp. 61. 1868.

BUCHENAU, F. *Luronium*, in Engler, H. G. A. (Ed.), *Das Pflanzenreich* **16** (IV. 15), 17–18. 1903.

GLÜCK, H. *Biologische und morphologische Untersuchungen über Wasser- und Sumpfgewächse*, 1. *Die Lebensgeschichte der europäischen Alismaceen*. Pp. xxiv + 312 + Taf. 7. Jena. 1905.

MICHELI, M. *Elisma*, in de Candolle, A. and de Candolle, C., *Monographie Phanerogamarum* **3**, 40–41. Paris. 1881.

ALISMA

BEAL, O. E. The Alismataceae of the Carolinas. *J. Elisha Mitchell Sci. Soc.* **76**, 68–79. 1960.

BUCHENAU, F. Index criticus Butomacearum, Alismacearum, Juncaginacearumque hucusque descriptarum. *Abh. Nat. Ver. Bremen* **1**. Pp. 61. 1868.

BUCHENAU, F. *Alisma*, in Engler, H. G. A. (Ed.), *Das Pflanzenreich* **16** (IV. 15), 12–15. 1903.

GLÜCK, H. *Biologische und morphologische Untersuchungen über Wasser- und Sumpfgewächse*, 1. *Die Lebensgeschichte der europäischen Alismaceen*. Pp. xxiv + 312 + Taf. 7. Jena. 1905.

HENDRICK, A. J. A revision of the genus Alisma (Dill.) L. *Amer. Midl. Nat.* **58**, 470–493. 1957. But see Voss, E. *Taxon* **7**, 130–133. 1958.

Micheli, M. *Alisma*, in de Candolle, A. and de Candolle, C., *Monographie Phanerogamarum* **3**, 31–38. Paris. 1881.

Pogan, E. Rangu systematyczna Alisma subcordatum Raf. i Alisma triviale Pursh. *Acta Biol. Cracov.* **6**, 185–202. 1963.

Samuelsson, G. Die Arten der Gattung Alisma L. *Arkiv Bot.* **24A** (7). Pp. 46 + Taf. 6. 1932.

DAMASONIUM

Beal, O. E. The Alismataceae of the Carolinas. *J. Elisha Mitchell Sci. Soc.* **76**, 68–79. 1960.

Buchenau, F. Index criticus Butomacearum, Alismacearum, Juncaginacearumque hucusque descriptarum. *Abh. Nat. Ver. Bremen* **1**. Pp. 61. 1868.

Buchenau, F. *Damasonium*, in Engler, H. G. A. (Ed.), *Das Pflanzenreich* **16** (IV. 15), 19–21. 1903.

Glück, H. *Biologische und morphologische Untersuchungen über Wasser- und Sumpfgewächse*, 1. *Die Lebensgeschichte der europäischen Alismaceen*. Pp. xxiv + 312 + Taf. 7. Jena. 1905.

Micheli, M. *Damasonium*, in de Candolle, A. and de Candolle, C., *Monographie Phanerogamarum* **3**, 41–44. Paris. 1881.

SAGITTARIA

Beal, O. E. The Alismataceae of the Carolinas. *J. Elisha Mitchell Sci. Soc.* **76**, 68–79. 1960.

Bogin, C. Revision of the genus Sagittaria (Alismataceae). *Mem. New York Bot. Gard.* **9**, 179–223. 1955.

Buchenau, F. Index criticus Butomacearum, Alismacearum, Juncaginacearumque hucusque descriptarum. *Abh. Nat. Ver. Bremen* **1**. Pp. 61. 1868.

Buchenau, F. *Sagittaria*, in Engler, H. G. A. (Ed.), *Das Pflanzenreich* **16** (IV. 15), 37–59. 1903.

Glück, H. *Biologische und morphologie Untersuchungen über Wasser- und Sumpfgewachse*, 1. *Die Lebensgeschichte der europäischen Alismaceen*. Pp. xxiv + 312 + Taf. 7. Jena. 1905.

Micheli, M. *Sagittaria*, in de Candolle, A. and de Candolle, C., *Monographie Phanerogamarum* **3**, 64–81. Paris. 1881.

Smith, J. G. North American species of Sagittaria and Lophotocarpus. *Ann. Rep. Missouri Bot. Gard.* **6**, 27–64 + 29 plates. 1895.

BUTOMACEAE

BUTOMUS

Buchenau, F. Index criticus Butomacearum, Alismacearum, Juncaginacearumque hucusque descriptarum. *Abh. Nat. Ver. Bremen* **1**. Pp. 61. 1868.

Buchenau, F. *Butomus*, in Engler, H. G. A. (Ed.), *Das Pflanzenreich* **16** (IV. 16), 5–6. 1903.

MICHELI, M. *Butomus*, in de Candolle, A. and de Candolle, C., *Monographie Phanerogamarum* **3**, 85–87. Paris. 1881.

HYDROCHARITACEAE

CASPARY, R. Conspectus systematicus Hydrillearum. *Monatsber. Koenigl. Akad. (Berlin)* **1857**. Pp. 15. 1857.

CASPARY, R. Die Hydrillen (Anacharideen Endl.). *Jahrb. Wiss. Bot.* **1**, 377–511 + Taf. 25–29. 1858.

GLÜCK, H. *Biologische und morphologische Untersuchengen über Wasser- und Sumpfgewächse* **2**. Pp. xvii + 256 + Taf. 6. Jena. 1906.

HYDROCHARIS

GLÜCK, H. *Biologische und morphologische Untersuchungen über Wasser- und Sumpfgewächse* **2**. Pp. xvii + 256 + Taf. 6. Jena. 1906.

STRATIOTES

GLÜCK, H. *Biologische und morphologische Untersuchungen über Wasser- und Sumpfgewächse* **2**. Pp. xvii + 256 + Taf. 6. Jena. 1906.

EGERIA

CASPARY, R. Conspectus systematicus Hydrillearum. *Monatsber. Koenigl. Akad. (Berlin)* **1857**. Pp. 15. 1857.

CASPARY, R. Die Hydrillen (Anacharideen Endl.). *Jahrb. Wiss. Bot.* **1**, 377–511 + Taf. 25–29. 1858.

ST. JOHN, H. Monograph of the genus Egeria Planchon. *Darwiniana* **12**, 293–307. 1961.

ELODEA

CASPARY, R. Conspectus systematicus Hydrillearum. *Monatsber. Koenigl. Akad. (Berlin)* **1857**. Pp. 15. 1857.

CASPARY, R. Die Hydrillen (Anacharideen Endl.). *Jahrb. Wiss. Bot.* **1**, 377–511 + Taf. 25–29. 1858.

GLÜCK, H. *Biologische und morphologische Untersuchungen über Wasser- und Sumpfgewächse* **2**. Pp. xvii + 256 + Taf. 6. Jena. 1906.

MARIE-VICTORIN, Frère. L'Anacharis canadensis, histoire et solution d'un imbroglio taxonomique. *Contr. Lab. Bot. Univ. Montréal* **18**, 7–43. 1931.

ST. JOHN, H. Monograph of the genus Elodea (Hydrocharitaceae), 1. The species found in the Great Plains, the Rocky Mountains and the Pacific States and Provinces of North America. *Res. Stud. Washington State Univ.* **30**, 19–44. 1962: 2. The species found in the Andes and western South America. *Caldasia* **9**, 95–113. 1964: 3. The species found in northern and eastern South America. *Darwiniana* **12**, 639–652. 1963: 4, and summary. The species of eastern and central North America. *Rhodora* **67**, 1–35, 155–180. 1965.

LAGAROSIPHON

Caspary, R. Die Hydrillen (Anacharideen Endl.) *Jahrb. Wiss. Bot.* **1**, 377–511 + Taf. 25–29. 1858.

Obermeyer, A. A. The South African species of Lagarosiphon. *Bothalia* **8**, 139–146. 1964.

VALLISNERIA

Marie-Victorin, Frère. Les Vallisnéries américaines. *Contr. Lab. Bot. Univ. Montréal* **46**, 1–38. 1943.

SCHEUCHZERIACEAE

SCHEUCHZERIA

Buchenau, F. *Scheuchzeria*, in Engler, H. G. A. (Ed.), *Das Pflanzenreich* **16** (IV. 14), 14–15. 1903.

Micheli, M. *Scheuchzeria*, in de Candolle, A. and de Candolle, C., *Monographie Phanerogamarum* **3**, 95–96. Paris. 1881.

JUNCAGINACEAE

TRIGLOCHIN

Buchenau, F. Index criticus Butomacearum, Alismacearum, Juncaginacearumque hucusque descriptarum. *Abh. Nat. Ver. Bremen* **1**. Pp. 61. 1868.

Buchenau, F. *Triglochin*, in Engler, H. G. A. (Ed.), *Das Pflanzenreich* **16** (IV. 14), 7–14. 1903.

Löve, A. and Löve, D. Biosystematics of Triglochin maritimum agg. *Nat. Canad.* **85**, 156–165. 1958.

Micheli, M. *Triglochin*, in de Candolle, A. and de Candolle, C., *Phanerogamarum* **3**, 96–109. Paris. 1881.

APONOGETONACEAE

APONOGETON

Camus, A. Le genre Aponogeton L.f. *Bull. Soc. Bot. France* **70**, 670–676. 1923.

Krause, K. and Engler, A. *Aponogeton*, in Engler, H. G. A. (Ed.), *Das Pflanzenreich* **24** (IV. 13), 9–23. 1906.

ZOSTERACEAE

ZOSTERA

Ascherson, P. and Graebner, P. *Zostera*, in Engler, H. G. A. (Ed.), *Das Pflanzenreich* **31** (IV. 11), 27–33. 1907.

Tutin, T. G. New species of Zostera from Britain. *J. Bot.* **74**, 227–230. 1936.

POTAMOGETONACEAE

POTAMOGETON

Ascherson, P. and Graebner, P. *Potamogeton*, in Engler, H. G. A. (Ed.), *Das Pflanzenreich* **31** (IV. 11), 39–142. 1907.

Dandy, J. E. and Taylor, G. Studies of British Potamogetons. *J. Bot.* **76**, 166–171, 239–241.1938: *loc. cit.* **77**, 56–62, 97–101, 161–164, 253–259, 277–282, 304–311.1939: *loc. cit.* **78**, 1–11, 50–66.1940: *loc. cit.* **80**, 117–142.1942.

Dandy, J. E. and Taylor, G. Two new British hybrid pondweeds. *Kew Bull.* **1957**, 332.1957.

Fryer, A. and Bennett, A. *The Potamogetons (Pond Weeds) of the British Isles.* Pp. 1–36. London. 1898: Pp. 37–56.1900: Pp. 57–76.1912: Pp. 77–94. 1915 + 60 plates.

Glück, H. *Biologische und morphologische Untersuchungen über Wasser- und Sumpfgewächse* 2. Pp. xvii + 256 + Taf. 6. Jena 1906: *loc. cit.* 4. Pp. viii + 746 + taf. 8. 1924.

RUPPIACEAE

RUPPIA

Ascherson, P. and Graebner, P. *Ruppia,* in Engler, H. G. A. (Ed.), *Das Pflanzenreich* **31** (IV.11), 142–145.1907.

Setchell, W. A. The genus Ruppia. *Proc. Calif. Acad. Sci. ser.* 4 **25** (18), 469–478.1946.

NAJADACEAE

NAJAS

Braun, A. Revision of genus Najas of Linnaeus. *J. Bot.* **2**, 274–279.1864.

Magnus, P. *Beiträge zur Kenntnis der Gattung Najas L.* Pp. 63. Berlin. 1870.

Rendle, A. B. The British species of Najas. *J. Bot.* **38**, 105–109.1900.

Rendle, A. B. *Najas,* in Engler, H. G. A. (Ed.), *Das Pflanzenreich* **7** (IV.12), 6–19.1901.

ERIOCAULACEAE

ERIOCAULON

*Koernicke, F. *Monographia scripta de Eriocaulaceis.* Pp. ii + 36. Berlin. 1826.

*Koernicke, F. Eriocaulacearum monographiae supplementum. *Linnaea* **27**, 561–692.1827.

Ruhland, W. *Eriocaulon,* in Engler, H. G. A. (Ed.), *Das Pflanzenreich* **13** (IV.30), 30–117.1903.

LILIACEAE

TOFELDIA

Smith, J. E. A botanical history of the genus Tofeldia. *Trans. Linn. Soc.* **12**, 235–247.1818.

POLYGONATUM

Ownbey, R. P. The Liliaceous genus Polygonatum in North America. *Ann. Missouri Bot. Gard.* **31**, 373–413.1944.

MAIANTHEMUM

INGRAM, J. Notes on the cultivated Liliaceae, 3. Maianthemum. *Baileya* **14**, 50–59. 1966.

ASPARAGUS

BOZZINI, A. Revisione cito-sistematica del genero Asparagus. *Caryologia* **12**, 199–264. 1959.

LILIUM

ELWES, H. J. *A Monograph of the genus Lilium.* Parts 1–7. London. 1877–80.

GROVE, A. and COTTON, A. D. *A Supplement to Elwes' Monograph of Lilium.* Parts 1–7. London. 1933–40.

TURRILL, W. B. *Second Supplement to Elwes' Monograph of Lilium.* Part 8. London. 1960. Part 9. 1962.

FRITILLARIA

BECK, C. *Fritillaries.* Pp. 96. London. 1953.

GROVE, A. The genus Fritillaria. *New Flora and Silva* **3**, 147–153. 1931.

ROSENHEIM, P. European species of the genus Fritillaria. *Quart. Bull. Alpine Gard. Soc.* **7**, 1–15, 110–124. 1939.

TULIPA

HALL, A. D. *The Genus Tulipa.* Pp. viii + 171 + 40 plates. London. 1940.

LEVIER, E. Les Tulipes de l'Europe. *Bull. Soc. Neuchât. Sci. Nat.* **14**, 201–312 + 10 plates. 1884.

GAGEA

STROH, G. Die Gattung Gagea Salisb. *Beih. Bot. Centr.* **57B**, 485–520. 1937: *loc. cit.* **58B**, 213–214. 1938.

UPHOF, J. C. T. A review of the genus Gagea Salisb. *Plant Life* **14**, 132–142. 1958: *loc. cit.* **14**, 151–161. 1959.

ORNITHOGALUM

BAKER, J. G. Revision of the genera and species of Scillae and Chlorogalae. *J. Linn. Soc. Bot.* **13**, 209–292. 1872–73.

ZAHARIADI, C. Sous-genres et sections mésogéens du genre Ornithogalum et la valeur comparative de leurs caractères différentials. *Rev. Roum. Biol. sér. Bot.* **10**, 271–291. 1965.

COLCHICUM

BOWLES, E. A. *A Handbook of Crocus and Colchicum for Gardeners.* Pp. xii + 185. London. 1924. Edition 2. Pp. 222. 1952.

D'AMATO, F. Revisione citosistematica del genere Colchicum. *Caryologia* **7**, 292–348. 1955: *loc. cit.* **9**, 315–339. 1956: *loc. cit.* **10**, 111–151. 1957.

STEFANOFF, B. Monografiya na roda Colchicum. *Sborn. Bălg. Akad. Nauk.* **22**. Pp. 100. 1926.

PARIS

*Franchet, A. Monographie du genre Paris. *Mém. Soc. Philom. (Paris)*. 1888.

JUNCACEAE

Buchenau, F. Monographie der Juncaceen vom Cap. *Abh. Nat. Ver. Bremen* **4**, 393–512. 1875.

Buchenau, F. Kritisches Verzeichness aller bis jetzt beschriebenen Juncaceen nebst Diagnosen neuer Arten. *Abh. Nat. Ver. Bremen* **13**. Pp. 112. 1880.

Buchenau, F. Die Verbreitung der Juncaceen über die Erde. *Bot. Jahrb.* **1**, 104–141. 1881.

Buchenau, F. Kritische Zusammenstellung der europäischen Juncaceen. *Bot. Jahrb.* **7**, 153–176. 1886.

Buchenau, F. Monographia Juncacearum. *Bot. Jahrb.* **12**, 1–495, 622. 1890.

Husnot, T. *Joncées. Descriptiones et Figures des Joncées de France, Suisse et Belgique.* Pp. 28 + 7 plates. Cahan. 1908.

JUNCUS

Bicheno, J. E. Observations on the Linnean genus Juncus with the characters of those species which have been found growing wild in Great Britain. *Trans. Linn. Soc.* **12**, 291–337. 1818.

Blau, J. *Vergleichend-anatomische Untersuchungen der Schweizerischen Juncus-Arten.* Pp. 78. Zürich. 1904.

Buchenau, F. Monographie der Juncaceen vom Cap. *Abh. Nat. Ver. Bremen* **4**, 393–512. 1875.

Buchenau, F. Kritisches Verzeichniss aller bis jetzt beschrieben Juncaceen nebst Diagnosen neuer Arten. *Abh. Nat. Ver. Bremen* **13**. Pp. 112. 1880.

Buchenau, F. Die Verbreitung der Juncaceen über die Erde. *Bot. Jahrb.* **1**, 104–141. 1881.

Buchenau, F. Kritische Zusammenstellung der europäischen Juncaceen. *Bot. Jahrb.* **7**, 153–176. 1886.

Buchenau, F. Monographia Juncacearum. *Bot. Jahrb.* **12**, 1–495, 622. 1890.

Buchenau, F. *Juncus*, in Engler, H. G. A. (Ed.), *Das Pflanzenreich* **25** (IV. 36), 98–266. 1906.

Edgar, E. The leafless species of Juncus in New Zealand. *N.Z.J. Bot.* **2**, 177–204. 1964.

Husnot, T. *Joncées. Descriptiones et Figures des Joncées de France, Suisse et Belgique.* Pp. 28 + 7 plates. Cahan. 1908.

Křisa, B. Taxonomische Stellung der Art Juncus alpinus Vill. s.l. in der europäischen flora. *Nov. Bot. Hort. Bot. Univ. Prag.* **1963**, 28–38. 1963.

Lindquist, B. Taxonomical remarks on Juncus alpinus Villars and some related species. *Bot. Not.* **1932**, 314–372. 1932.

Meyer, E. *Junci Generis Monographiae Specimen.* Pp. 48. Gottingae. 1819.

MEYER, E. *Synopsis Juncorum rite cognitarum*. Pp. 66. Gottingae. 1822.
ROSTKOVIUS, F. W. G. *Dissertatio botanico de Junco*. Halae. 1801. Edition 2. *Monographia generis Junci*. Pp. 58. Berolini. 1801.

LUZULA

BUCHENAU, F. Monographie der Juncaceen vom Cap. *Abh. Nat. Ver. Bremen* **4**, 393–512. 1875.
BUCHENAU, F. Kritisches Verzeichniss aller bis jetzt beschrieben Juncaceen nebst Diagnosen neuer Arten. *Abh. Nat. Ver. Bremen* **13**. Pp. 112. 1880.
BUCHENAU, F. Die Verbreitung der Juncaceen über die Erde. *Bot. Jahrb.* **1**, 104–141. 1881.
BUCHENAU, F. Kritische Zusammenstellung der europäischen Juncaceen. *Bot. Jahrb.* **7**, 153–176. 1886.
BUCHENAU, F. Monographia Juncacearum. *Bot. Jahrb.* **12**, 1–495, 622. 1890.
BUCHENAU, F. *Luzula*, in Engler, H. G. A. (Ed.), *Das Pflanzenreich* **25** (IV. 36), 42–48. 1906.
CHRTEK, J. and KŘISA, B. A taxonomical study of the species Luzula spicata (L.) DC. sensu lato in Europe. *Bot. Not.* **115**, 293–310. 1962.
CHRTEK, J. and KŘISA, B. Notes on the Luzula spicata complex in Turkey. *Notes Roy. Bot. Gard. Edinb.* **25**, 163–164. 1964.
CHRTEK, J. and KŘISA, B. Bemerkungen zu den mediterraneen Arten Luzula spicata-Komplex. *Nov. Bot. Hort. Univ. Prag.* **1965**, 27–29. 1965.
EBINGER, J. E. Taxonomy of the subgenus Pterodes of genus Luzula. *Mem. New York Bot. Gard.* **10** (5), 279–304. 1964.
EDGAR, E. Luzula in New Zealand. *N.Z.J. Bot.* **4**, 159–184. 1966.
HUSNOT, T. *Joncées. Descriptiones et Figures des Joncées de France, Suisse et Belgique.* Pp. 28 + 7 plates. Cahan. 1908.
MEYER, E. H. F. *Synopsis Luzularum rite cognitorum*. Pp. viii + 40. Gottingae. 1823.
MONTSERRAT, P. El género Luzula en España. *Anal. Inst. Bot. Cav.* **21**, 407–523. 1963.

AMARYLLIDACEAE

BAKER, J. G. *Handbook of the Amaryllideae*. Pp. xii + 216. London. 1888.
HERBERT, W. *Amaryllidaceae*. Pp. 428 + 48 plates. London. 1837.

ALLIUM

DE JANKE, V. Key to the Alliums of Europe. *Herbertia* **11**, 219–225. 1946.
DON, G. A monograph of the genus Allium. *Mem. Wern. Nat. Hist. Soc.* **6**, 1–102. 1827.
FEINBRUN, N. Allium sectio Porrum of Palestine and the neighbouring countries. *Palestine J. Bot.* **3**, 1–21. 1943.
FEINBRUN, N. Further studies on Allium of Palestine and the neighbouring countries. *Palestine J. Bot.* **4**, 144–157. 1948.

Helm, J. Die zu Wuerz- und Spesiezwecken kultivierten Arten der Gattung Allium L. *Kulturpflanze* **4**, 130–180. 1956.

Moore, H. E., Jnr. The cultivated Alliums. *Baileya* **2**, 103–122. 1954: *loc. cit.* **3**, 137–149, 156–167. 1955.

Regel, E. Alliorum adhuc cognitorum Monographia. *Acta Hort. Petrop.* **3** (2), 1–266. 1875.

Regel, E. *Allii Species Asiae Centralis in Asia Media.* Pp. 87 + Tab. 8. Petropoli. 1887.

Vindt, J. Synopsis du genre Allium au Maroc. *Bull. Soc. Sci. Nat. Maroc.* **33**, 109–128. 1953.

Vredensky, A. I. *Allium*, in Komarov, V. L., *et al.* (Eds), *Flora URSS* **4**, 112–280. 1935. Translated into English by H. K. Airy Shaw. *Herbertia* **11**, 65–218. 1946.

LEUCOJUM

Baker, J. G. *Handbook of the Amaryllideae,* 18–20. London. 1888.

Stern, F. C. *Snowdrops and Snowflakes.* Pp. 128. London. 1956.

GALANTHUS

Artiushenko, Z. T. K sistematike roda Galanthus L. *Bot. Žurn.* **50**, 1430–1447. 1965.

Baker, J. G. *Handbook of the Amaryllideae,* 16–18. London. 1888.

Bowles, E. A. Snowdrops. *J. Roy. Hort. Soc.* **43**, 28–36. 1918.

Gottlieb-Tannenhain, P. Studien über die Formen der Gattung Galanthus. *Abh. Zool.-Bot. Ges. Wien* **2**. Pp. 95. 1904.

Stern, F. C. *Snowdrops and Snowflakes.* Pp. 128. London. 1956.

STERNBERGIA

Baker, J. G. *Handbook of the Amaryllideae,* 28–29. London. 1888.

NARCISSUS

Baker, J. G. *Handbook of the Amaryllideae,* 3–14. London. 1888.

Burbidge, F. W. *The Narcissus: its History and Culture, with coloured plates and descriptions of all known species and principal varieties to which is added by kind permission, a scientific review of the entire genus by J. G. Baker, F.L.S.* Pp. 95 + 48 plates. London. 1875.

Fernandes, A. Sur la caryo-systématique du groupe Jonquilla du genre Narcissus L. *Bol. Soc. Brot. sér.* 2 **13**, 487–544. 1939.

Fernandes, A. Sur la phylogénie des espèces du genre Narcissus L. *Bol. Soc. Brot. sér.* 2 **25**, 113–190. 1951.

Fernandes, A. Nouvelles études caryologiques sur la section Jonquilla DC. du genre Narcissus L. *Bol. Soc. Brot. sér.* 2 **40**, 207–261. 1966.

Haworth, A. *Narcissineäurum Monographia.* Pp. vi + 7–23. London. 1831.

Herbert, W. *Amaryllidaceae*, 316–318. London. 1837.

Pugsley, H. W. Narcissus poeticus and its allies. *J. Bot.* **53** Supplement 2, 1–44 + 2 plates. 1915.

Pugsley, H. W. A monograph of Narcissus sub-genus Ajax. *J. Roy. Hort. Soc.* **58**, 17–93. 1933.

IRIDACEAE

Baker, J. G. *Handbook of Irideae.* Pp. xii + 247. London. 1892.

SISYRINCHIUM

Baker, J. G. *Handbook of Irideae*, 121–133. London. 1892.

Bicknell, E. Studies in Sisyrinchium. *Bull. Torrey Bot. Club* **26**, 217–231, 297–300, 335–349, 445–457, 496–499, 605–616. 1899: *loc. cit.* **27**, 237–246, 373–387. 1900: *loc. cit.* **28**, 570–592. 1901: *loc. cit.* **31**, 379–391. 1904.

IRIS

Anderson, E. The species problem in Iris. *Ann. Missouri Bot. Gard.* **23**, 457–509. 1936.

Baker, J. G. *Handbook of Irideae*, 1–47. London. 1892.

Dykes, W. R. *The Genus Iris.* Pp. iii + 245. Cambridge. 1913.

Foster, R. C. A cytotaxonomic survey of the North American species of Iris. *Contr. Gray Herb.* **119**, 3–80. 1937.

Lawrence, G. H. M. A reclassification of the genus Iris. *Gentes Herb.* **8**, 345–371. 1953.

Rodionenko, G. I. *Rod Iris L.* Pp. 216. Moskva and Leningrad. 1961.

Spach, E. Revisio Generis Iris. *Ann. Sci. Nat. sér.* 3 *Bot.* **5**, 89–111. 1846.

HERMODACTYLUS

Baker, J. G. *Handbook of Irideae*, 47–48. London. 1892.

Spach, E. Revisio Generis Iris. *Ann. Sci. Nat. sér.* 3 *Bot.* **5**, 89–111. 1846.

CROCUS

Baker, J. G. *Handbook of Irideae*, 76–95. London. 1892.

Bowles, E. A. *A Handbook of Crocus and Colchicum for Gardeners.* Pp. iii + 185 London. 1924. Edition 2. Pp. 222. 1952.

Lawrence, G. H. M. Keys to cultivated plants, 4. The spring-flowering crocuses. *Baileya* **3**, 127–137. 1954.

Maw, G. A. *Monograph of the genus Crocus.* Pp. xx + 326. London. 1886.

ROMULEA

Baker, J. G. *Handbook of Irideae*, 97–104. London. 1892.

Béguinot, A. Revisione monografica del genero Romulea Maratti. *Malpighia* **21**, 49–122, 364–378. 1907: *loc. cit.* **22**, 377–469. 1908: *loc. cit.* **23**, 55–117, 185 bis–239, 275–293. 1909.

MARIN, A. Contribution à l'étude des Romulea de la Côte Nord-Atlantique Marocaine. *Trav. Inst. Sci. Chérif. sér. Bot.* **27**, 7–48. 1962.

GLADIOLUS

BAKER, J. G. *Handbook of Irideae*, 198–229. London. 1892.

DIETRICH, K. F. *Ueber die europäischen Arten der Gattung Gladiolus*. Pp. 13. Berlin. 1832.

KLATT, F. W. Revisio Iridearum. *Linneana* **32**, 689–801. 1863.

SYME, J. T. B. Remarks on Gladiolus illyricus and its allies. *J. Bot.* **1**, 130–134. 1863.

CYPRIPEDIACEAE (See also ORCHIDACEAE)

CYPRIPEDIUM

DESBOIS, F. *Monographie des Cypripedium, Selepodium et Uropedium*. Pp. 159 Gand. 1888. Edition 2. Pp. 544. 1898.

KRAENZLIN, F. *Orchidacearum Genera et Species*, 1. *Apostasieae*, *Cypripediaceae*, *Ophrydeae*, 11–86. Berlin. 1897.

ORCHIDACEAE

CAMUS, E. G. *Monographie des Orchidées de l'Europe, de l'Afrique septentrionale de le Asie Mineure et des Provinces Russes transcaspiennes*. Pp. 484 + 32 plates. Paris. 1908.

CAMUS, E. G. and CAMUS, A. *Iconographie des Orchidées d'Europe et du bassin méditerrannéen*. Pp. 559 + 133 plates. Paris. 1928–29.

DANESCH, O. and DANESCH, E. *Orchideen Europas: Mitteleuropas*. Pp. 264 + 165 illustrations. Stuttgart. 1963.

DUPERREX, A. and DOUGOUD, R. *Orchidées d'Europe*. Pp. 239. Neuchâtel and Paris. 1955. Translated into English by A. J. Huxley, Pp. xiii + 15–235. London. 1961.

GODFERY, M. J. *Monograph and Iconograph of Native British Orchidaceae*. Pp. xiv + 259. Cambridge. 1933.

KELLER, R., SCHLECHTER, R. and SOÓ, R. von. Monographie und Iconographie der Orchideen Europas und des Mittelmeergebietes. *Fedde. Rep. Sonderbeih.* **A1**. Pp. 304. 1925–28: 2. Pp. 472. 1930–40: 3. Plates 1–96. 1931–33: 3 (2). Plates 97–192. 1933–35: 4. Plates 193–288. 1935–36: 4 (2). Plates 289–400. 1937–38: 5. Plates 401–499. 1939–43: 5 (2). Plates 500–640. 1939–43.

KRAENZLIN, F. *Orchidacearum Genera et Species*, 1. *Apostasieae*, *Cypripedieae*, *Ophrydeae*. Pp. viii + 986. Berlin. 1897–1901.

SCHUSTER, C. Orchidacearum Iconum index Zusammenstellung der in der Literatur erscheinenden Tafeln und Textbildungen von Orchideen. *Fedde. Rep. Beih.* **60**, 1–80. 1931: *loc. cit.* **60**, 81–160. 1932: *loc. cit.* **60**, 161–320. 1933: *loc. cit.* **60**, 321–400. 1934: *loc. cit.* **60**, 401–536. 1936.

SUMMERHAYES, V. S. *Wild Orchids of Britain*. Pp. xvii + 366. London. 1951.
TAHOURDIN, C. B. *Native Orchids of Britain*. Pp. 114. Croydon. 1925.
WEBSTER, A. D. *British Orchids*. London. 1886. Edition 2. Pp. viii + 131. 1898.

EPIPACTIS

NANNFELDT, J. A. Tre för Norden nya Epipactis-arter, E. persica Hausskn., E. leptochila (Godf.) Godf. och E. purpurata Sm. *Bot. Not.* **1946**, 1–28.1946.
STEPHENSON, T. and STEPHENSON, T. A. The genus Epipactis in Britain. *J. Bot.* **58**, 209–213.1920.
YOUNG, D. P. Studies in British Epipactis. *Watsonia* **1**, 102–113.1949: *loc. cit.* **2**, 253–276.1951: *loc. cit.* **5**, 127–142.1962.
YOUNG, D. P. Autogamous Epipactis in Scandinavia. *Bot. Not.* **1953**, 253–270.1953.
YOUNG, D. P. A key to the Danish Epipactis. *Bot. Tidsskr.* **50**, 140–145.1954.
YOUNG, D. P. Le genre Epipactis en Belgique. *Bull. Jard. Bot. Bruxelles* **28**, 123–127.1958.
YOUNG, D. P. and RENZ, J. Epipactis leptochila Godf.-its occurrence in Switzerland and its relationship to other Epipactis species. *Bauhinia* **1**, 151–156.1958.

EPIPOGIUM

SCHLECHTER, R. Die Polychondreae (Neottiinae Pfitz.) und ihre systematische Einteilung. *Bot. Jahrb.* **45**, 375–410.1911.

LIPARIS

RIDLEY, H. N. A monograph of the genus Liparis. *J. Linn. Soc. Bot.* **22**, 244–297.1886: *loc. cit.* **24**, 349–350.1888.

HERMINIUM

KRAENZLIN, F. *Orchidacearum Genera et Species*, 1. *Apostasieae, Cypripedieae, Ophrydeae* **1**, 530–536. Berlin. 1898.

COELOGLOSSUM

KRAENZLIN, F. Beiträge zur einer Monographie der Gattung Habenaria Willd. 2, Systematischer Teil. *Bot. Jahrb.* **16**, 52–223.1893.
KRAENZLIN, F. *Orchidacearum Genera et Species*, 1. *Apostasieae, Cypripedieae, Ophrydeae* **1**, 613–618. Berlin. 1898.

GYMNADENIA

KRAENZLIN, F. Beiträge zur einer Monographie der Gattung Habenaria Willd. 2, Systematischer Teil. *Bot. Jahrb.* **16**, 52–223.1893.

KRAENZLIN, F. *Orchidacearum Genera et Species*, 1. *Apostasieae, Cypripedieae, Ophrydeae* **1**, 550–565. Berlin. 1898.

PSEUDORCHIS (***LEUCORCHIS***)

KRAENZLIN, F. Beiträge zur einer Monographie der Gattung Habenaria Willd. 2, Systematischer Teil. *Bot. Jahrb.* **16**, 52–223. 1893.

KRAENZLIN, F. *Orchidacearum Genera et Species*, 1. *Apostasieae, Cypripedieae, Ophrydeae* **1**, 554–555. Berlin. 1898.

PLATANTHERA

KRAENZLIN, F. Beiträge zur einer Monographie der Gattung Habenaria Willd. 2, Systematischer Teil. *Bot. Jahrb.* **16**, 52–223. 1893.

KRAENZLIN, F. *Orchidacearum Genera et Species*, 1. *Apostasieae, Cypripedieae, Ophrydeae* **1**, 601–648. Berlin. 1898.

NEOTINEA

KRAENZLIN, F. *Orchidacearum Genera et Species* 1, *Apostasieae, Cypripedieae, Ophrydeae*, 172–174. Berlin. 1897.

OPHRYS

GODFERY, M. J. The genus Ophrys. *J. Bot.* **55**, 329–334. 1917.

KRAENZLIN, F. *Orchidacearum Genera et Species* 1, *Apostasieae, Cypripedieae, Ophrydeae*, 89–109. Berlin. 1897.

NELSON, E. *Gestaltwandel und Artbildung erörtert am Beispel der Orchidaceen Europas und der Mittelmeerländer, insbesondere der Gattung Ophrys, mit einer Monographie und Ikonographie der Gattung Ophrys.* Pp. 250 + 50 plates + 8 maps. Chernex-Montreux. 1962.

SOÓ, R. Ophrys-studien. *Acta Bot. Hung.* **5**, 437–471. 1959.

HIMANTOGLOSSUM

KRAENZLIN, F. *Orchidacearum Genera et Species* 1, *Apostasieae, Cypripedieae, Ophrydeae*, 164–168. Berlin. 1897.

ORCHIS

KRAENZLIN, F. *Orchidacearum Genera et Species* 1, *Apostasieae, Cypripedieae, Ophrydeae*, 40–144. Berlin. 1897.

DACTYLORHIZA (***DACTYLORCHIS***)

HESLOP-HARRISON, J. A synopsis of the Dactylorchids of the British Isles. *Veröff. Geobot. Inst. Rübel* (*Zürich*) **1953**, 53–82. 1954.

KRAENZLIN, F. *Orchidacearum Genera et Species* 1, *Apostasieae, Cypripedieae, Ophrydeae*, 144–154. Berlin. 1897.

PUGSLEY, H. W. New British Marsh Orchids. *Proc. Linn. Soc.* **148**, 121–125. 1936.

Soó, R. Synopsis generis Dactylorhiza (Dactylorchis). *Ann. Univ. Sci. Budapest.* **3**, 335–357. 1960.

Stephenson, T. and Stephenson, T. A. A new marsh orchis. *J. Bot.* **58**, 164–170. 1920.

Stephenson, T. and Stephenson, T. A. The British Palmate Orchids. *J. Bot.* **58**, 257–262. 1920.

Vermeulen, P. *Studies on Dactylorchids.* Pp. 180. Utrecht. 1947.

Wilmott, A. J. New British Marsh Orchids. *Proc. Linn. Soc.* **148**, 126–130. 1936.

ACERAS

Kraenzlin, F. *Orchidacearum Genera et Species* 1, *Apostasieae, Cypripedieae, Ophrydeae*, 164–168. Berlin. 1897.

ANACAMPTIS

Kraenzlin, F. *Orchidacearum Genera et Species* 1, *Apostasieae, Cypripedieae Ophrydeae*, 168–172. Berlin. 1897.

ARACEAE

ACORUS

Engler, A. *Acorus*, in de Candolle, A. and de Candolle, C., *Monographie Phanerogamarum* **2**, 215–218. Paris. 1879.

Engler, A. *Acorus*, in Engler, H. G. A. (Ed.), *Das Pflanzenreich* **21** (IV. 23B), 309–313. 1905.

Wein, K. Die älteste Einführungs -und Ausbreitungsgeschichte von Acorus calamus. *Hercynia* **1**, 367–450. 1939: *loc. cit.* **3**, 7–128. 1940: *loc. cit.* **3**, 241–291. 1948.

CALLA

Krause, K. *Calla*, in Engler, H. G. A. (Ed.), *Das Pflanzenreich* **37** (IV. 23B), 154–155. 1908.

LYSICHITON

Hultén, E. and St. John, H. The American species of Lysichitum. *Svensk Bot. Tidskr.* **25**, 453–464. 1931.

Krause, K. *Lysichiton*, in Engler, H. G. A. (Ed.), *Das Pflanzenreich* **37** (IV. 23B), 148–150. 1908.

ARUM

Engler, A. *Arum*, in de Candolle, A. and de Candolle, C., *Monographie Phanerogamarum* **2**, 580–597. Paris. 1879.

Engler, A. *Arum*, in Engler, H. G. A. (Ed.), *Das Pflanzenreich* **73** (IV. 23F), 67–99. 1920.

HRUBY, J. The genus Arum. *Bull. Soc. Bot. Genève* **4**, 113–160, 330–371. 1912.
PRIME, C. T. *Lords and Ladies*. Pp. xiv + 241. London. 1960.

LEMNACEAE

DAUBS, E. H. A monograph of Lemnaceae. *Illinois Biol. Monogr.* **34**, 1–118 + 21 plates. 1965.
HEGELMAIER, F. Systematische Übersicht der Lemnaceen. *Bot. Jahrb.* **21**, 268–305. 1896.
SCHLEIDEN, M. J. Prodromus monographiae Lemnacearum oder Conspectus generum atque specierum. *Linnaea* **13**, 385–392. 1839.

LEMNA

DAUBS, E. H. A monograph of Lemnaceae. *Illinois Biol. Monogr.* **34**, 1–118 + 21 plates. 1965.
HEGELMAIER, F. Systematische Übersicht der Lemnaceen. *Bot. Jahrb.* **21**, 268–305. 1896.
SCHLEIDEN, M. J. Prodromus monographiae Lemnacearum oder Conspectus generum atque specierum. *Linnaea* **13**, 385–392. 1839.
WOLFF, J. F. *Commentatio de Lemna*. Pp. 22. Norimbergae. 1807.

SPIRODELA (*LEMNA*, p.p.)

DAUBS, E. H. A monograph of Lemnaceae. *Illinois Biol. Monogr.* **34**, 1–118 + 21 plates. 1965.
HEGELMAIER, F. Systematische Übersicht der Lemnaceen. *Bot. Jahrb.* **21**, 268–305. 1896.
SCHLEIDEN, M. J. Prodromus monographiae Lemnacearum oder Conspectus generum atque specierum. *Linnaea* **13**, 385–392. 1839.
WOLFF, J. F. *Commentatio de Lemna*. Pp. 22. Norimbergae. 1807.

WOLFFIA

DAUBS, E. H. A monograph of Lemnaceae. *Illinois Biol. Monogr.* **34**, 1–118 + 21 plates. 1965.
HEGELMAIER, F. Systematische Übersicht der Lemnaceen. *Bot. Jahrb.* **21**, 268–305. 1896.
SCHLEIDEN, M. J. Prodromus monographiae Lemnacearum oder Conspectus generum atque specerium. *Linnaea* **13**, 385–392. 1839.
WOLFF, J. F. *Commentatio de Lemna*. Pp. 22. Norimbergiae. 1807.

SPARGANIACEAE

SPARGANIUM

BEAL, E. O. Sparganium in the south eastern United States. *Brittonia* **12**, 176–181. 1960.
COOK, C. D. K. Sparganium in Britain. *Watsonia* **5**, 1–10. 1961.

FERNALD, M. L. Notes on Sparganium. *Rhodora* **24**, 26–34. 1922.

GRAEBNER, P. *Sparganium*, in Engler, H. G. A. (Ed.), *Das Pflanzenreich* **2** (IV.9), 10–24. 1900.

TYPHACEAE

TYPHA

GÈZE, J. B. *Études botaniques et agronomiques sur les Typha et quelques autres plantes palustres*. Pp. viii + 174. Villefranche-de-Rouergue. 1912.

GRAEBNER, P. *Typha*, in Engler, H. G. A. (Ed.), *Das Pflanzenreich* **2** (IV.8), 8–16. 1900.

HOTCHKISS, N. and DOZIER, H. L. Taxonomy and distribution of North American cat-tails. *Amer. Midl. Nat.* **41**, 237–254. 1949.

KRONFIELD, E. M. Monographie der Gattung Typha Tourn. *Verh. Zool.-Bot. Ges. Wien* **39**, 89–192. 1889.

CYPERACEAE

ERIOPHORUM

FAEGRI, K. Zur Hybridbildung in der Gattung Eriophorum, in *Festschrift Werner Lüdi* (Bot. Inst. Rübel, Zürich), 50–58. 1958.

NYLANDER, F. Eriophori Monographia. *Acta Soc. Bot. Fenn.* **3**, 1–23. 1846.

SCIRPUS

BAKKER, D. Miscellaneous notes on Scirpus lacustris L. sensu lato in the Netherlands. *Acta Bot. Neerl.* **3**, 426–445. 1954.

BEETLE, A. Annotated list of original descriptions in Scirpus. *Amer. Midl. Nat.* **41**, 453–493. 1949.

KOYAMA, T. Taxonomic study of the genus Scirpus Linné. *J. Fac. Sci. Tokyo Univ. (Bot.)* **7**, 271–366. 1958.

KOYAMA, T. The genus Scirpus Linn. Critical species of the section Pterolepis. *Canad. J. Bot.* **41**, 1107–1131. 1963.

LOUSLEY, J. E. The Schoenoplectus group of the genus Scirpus in Britain. *J. Bot.* **69**, 151–163. 1931.

ELEOCHARIS

CINZERLING, J. D. *Eleocharis*, in Komarov, V. L., *et al.* (Eds), *Fl. URSS* **3**, 63–90, 580–587. 1935.

FERNALD, M. L. and BRACKETT, A. E. The representatives of Eleocharis palustris in North America. *Rhodora* **31**, 57–77. 1929.

STRANDHEDE, S. O. Eleocharis palustre in Scandinavia and Finland. *Bot. Not.* **114**, 416–434. 1961.

SVENSON, H. K. Monographic studies in the genus Eleocharis. *Rhodora* **31**, 121–135, 152–163, 167–191, 199–219, 224–242. 1929: *loc. cit.* **34**, 193–203, 215–227. 1932: *loc. cit.* **36**, 377–389. 1934: *loc. cit.* **39**, 210–231, 236–273. 1937: *loc. cit.* **41**, 1–19, 43–77, 90–110. 1939.

WALTERS, S. M. Eleocharis R. Br. (Biological Flora). *J. Ecol.* **37**, 192–206.1949.

ZUKOWSKI, W. Rodzaj Eleocharis R. Br. w Polsce. *Prace Kom. Biol.* (*Poznań*) **30** (2), 1–113.1965.

CYPERUS

KÜKENTHAL, G. *Cyperus*, in Engler, H. G. A. (Ed.), *Das Pflanzenreich* **101** (IV.20 (2)), 41–320.1935: *loc. cit.* **101** (IV.20 (3)), 402–448.1936: *loc. cit.* **101** (IV.20 (4)), 481–566.1936.

SCHOENUS

KÜKENTHAL, G. Vorarbeiten zu einer Monographie de Rhynchosporoideae. *Fedde. Rep.* **44**, 1–32, 65–101, 161–195.1938.

RHYNCHOSPORA

KÜKENTHAL, G. Vorarbeiten zu einer Monographie de Rhynchosporoideae: Rhynchospora. *Bot. Jahrb.* **74**, 375–509.1949: *loc. cit.* **75**, 90–195.1950: *loc. cit.* **75**, 273–314.1951.

CLADIUM

KÜKENTHAL, G. Vorarbeiten zu einer Monographie de Rhynchosporoideae. *Fedde. Rep.* **51**, 1–7, 139–193.1942.

CAREX

BOOTT, F. *Illustrations of the genus Carex.* Vol. 1. Pp. xii + 1–74 + plates 1–200. London. 1858: Vol. 2. Pp. iv + 75–104 + plates 201–310.1860: Vol. 3. Pp. iv + 105–126 + plates 311–411.1862: Vol. 4. Pp. 127–233 + plates 412–460.1867.

CHRIST, H. List of European Carices. *J. Bot.* **28**, 260–266.1885.

DAVIES, E. W. Notes on Carex flava and its allies. *Watsonia* **3**, 66–84.1953.

GOODENOUGH, S. Observations on the British species of Carex. *Trans. Linn. Soc.* **2**, 126–211.1794.

GOODENOUGH, S. Additional observations on the British species of Carex. *Trans. Linn. Soc.* **3**, 76–79.1791.

HOPPE, D. H. *Caricologia germanica; oder, Aufzählung der in Deutschland wildwachsenden Riedgräser.* Pp. viii + 104. Leipzig. 1826.

KÜKENTHAL, G. *Carex*, in Engler, H. G. A. (Ed.), *Das Pflanzenreich* **38** (IV.20), 67–824.1909.

MACKENZIE, K. K. and CREUTZBURG, H. C. *North American Cariceae.* 539 plates. New York. 1940.

NELMES, E. Notes on British Carices. *J. Bot.* **77**, 112–114, 259–266, 301–304.1939: *loc. cit.* **80**, 105–112.1942: *Rep. Bot. Soc. & E.C.* **13**, 334–337.1948: *Watsonia* **2**, 249–252.1952: *Not. Syst.* (*Leningrad*) **19**, 75–78.1959.

NELMES, E. Two critical groups of British sedges. *Rep. Bot. Soc. & E.C.* **13**, 95–105. 1947.

NELMES, E. and SPRAGUE, T. Notes on British Carices. *J. Bot.* **77**, 152–154, 179–181. 1939.

NEUMANN, A. Vorläufiger Bestimmungs-Schlüssel für Carex-Arten Nordwestdeutschlands in blütenlosen Zustande. *Mitt. Fl.-Soz. Arbetsgeitem.* **3**, 44–77. 1952.

SCHKUHR, C. *Beschreibung und Abbildungen von Riedgrasen.* Pp. 128 + tab. 93. Wittenberg. 1801. Edition 2. Pp. 94 + tab. 93. 1806. Supplement by C. Kunze. Pp. 206 + tab. 50. Leipzig 1840–50. The first edition was reprinted in French as *Histoire des Carex.* Pp. 195 + tab. 93. 1802.

SENAY, P. Le groupe des Carex flava et Carex oederi. *Bull. Mus. Hist. Nat. (Paris)* **22**, 618–624, 790–796. 1950: *loc. cit.* **23**, 146–152. 1951.

SMITH, J. E. Descriptions of five new British species of Carex. *Trans. Linn. Soc.* **5**, 264–273. 1800.

VICIOSO, C. Estudio monográfico sobre el género Carex in España. *Bol. Inst. Forest. Invest. Exp.* **30**(79). Pp.205. 1959.

WAHLENBERG, G. A monograph of the genus Carex. *Ann. Bot. (London)* **2**, 112–144. 1806.

GRAMINEAE

HITCHCOCK, A. S. *Manual of the Grasses of the United States.* Pp. 1040. *U.S. Dept. Agric. Misc. Publ.* **200**. 1935. Edition 2. Revised by A. Chase. Pp. 1051. 1950.

HUBBARD, C. E. *Grasses.* Pp. xii + 428. Harmondsworth. 1954.

PARNELL, R. *The Grasses of Britain.* Pp. 302 + 142 plates. 3 vols. London. 1842–45.

TRINIUS, C. B. *Species Graminum Iconibus et Descriptionibus.* 3 vols. Petropoli. 1828–36.

ARUNDINEAE

CONERT, H. J. *Die Systematik und Anatomie der Arundineae.* Pp. 208. Weinheim. 1961.

MOLINIA

MATUSZKIEWICZ, A. and MATUSZKIEWICZ, W. A contribution to the taxonomy of the genus Molinia Schrk. *Ann. Univ. Mariae-Curie* **3**, 347–367. 1948.

GLYCERIA

BORRILL, M. A biosystematic study of some Glyceria species in Britain. *Watsonia* **3**, 291–306. 1956: *loc. cit.* **4**, 77–100. 1958.

LAMBERT, J. L. The British Species of Glyceria, with a Key, in Wilmott, A. J. (Ed.), *British Flowering Plants and Modern Systematic Methods*, 86–89. London. 1951.

FESTUCA

DE LITARDIÈRE, R. Revision du groupe Festuca ovina L. subsp. alpina Hack. *Bull. Soc. Bot. France* **70**, 287–293. 1923.

DE LITARDIÈRE, R. Contribution à l'étude des Festuca (Subgen. Eu-Festuca) du Nord de la France (Nord, Pas-de-Calais) et de Belgique. *Bull. Soc. Bot. Belg.* **55**, 92–133, 149–154. 1923.

DE LITARDIÈRE, R. Contribution à l'étude du genre Festuca. *Candollea* **10**, 103–146. 1945.

DE LITARDIÈRE, R. Contribution à l'étude des Festuca du Portugal. *Agron. Lusit.* **14**, 31–51. 1952.

DE WILDE DUYFJES, B. E. E. Festuca ovina L., s.l. en Festuca rubra L., s.l. in Nederland. *Gorteria* **2**, 40–48. 1964.

HACKEL, E. *Monographia Festucarum europaearum*. Pp. 216. Kassel and Berlin. 1882.

HOLMEN, K. Cytotaxonomical studies in the Arctic Alaskan flora. The genus Festuca. *Bot. Not.* **117**, 109–119. 1964.

HOWARTH, W. O. On the occurrence and distribution of Festuca rubra, Hack. in Great Britain. *J. Linn. Soc. Bot.* **46**, 313–331 + 5 plates. 1923.

HOWARTH, W. O. On the occurrence and distribution of Festuca ovina L., sensu ampliss. in Britain. *J. Linn. Soc. Bot.* **47**, 29–39. 1925.

HOWARTH, W. O. The genus Festuca in New Zealand. *J. Linn. Soc. Bot.* **48**, 57–77. 1928.

HOWARTH, W. O. A synopsis of the British Fescues. *Rep. Bot. Soc. & E.C.* **13**, 338–346. 1948.

HUON, A. Revision des Festuca de l'Herbier Lloyd. *Bull. Soc. Sci. Anjou* **4**, 45–65. 1961.

MARKGRAF-DANNENBERG, I. Die Gattung Festuca in den Bayerischen Alpen. *Ber. Bayer. Bot. Ges.* **28**, 195–209. 1950.

MARKGRAF-DANNENBERG, I. Studien an irischen Festuca-Rassen. *Veröff. Geobot. Inst. Rübel* **25**, 114–142. 1952.

NYÁRÁDY, E. I. and NYÁRÁDY, A. Studie über die Arten der Sektion Ovinae Fr. der Gattung Festuca in der RVR. *Rev. Roum. Biol.* **9**, 99–137, 151–172. 1964.

RAUSCHERT, S. Studien über die Systematik und Verbreitung der thüringischen Sippen der Festuca ovina L.s. lat. *Fedde. Rep.* **63**, 251–283. 1960.

SAINT-YVES, A. Les Festuca (subg. Eu-Festuca) de l'Afrique du Nord et des Iles Atlantiques. *Candollea* **1**, 1–63. 1922.

SAINT-YVES, A. Contribution à l'étude des Festuca (Subgen. Eu-Festuca) de l'Amérique du Nord et du Mexique. *Candollea* **2**, 229–316. 1925.

SAINT-YVES, A. Contribution à l'étude des Festuca (Subgen. Eu-Festuca) de l'Amérique de Sud. *Candollea* **3**, 151–315. 1927.

SAINT-YVES, A. Contribution à l'étude des Festuca (Subgen. Eu-Festuca) de l'Orient, Asie et région méditerranéenne voisine. *Candollea* **3**, 320–466. 1928.

SAINT-YVES, A. Contribution à l'étude des Festuca (Subgen. Eu-Festuca) de l'Afrique australe et de l'Océanie. *Candollea* **4**, 65–129. 1929.

SAINT-YVES, A. Festuca de la Nouvelle Zélande (Herbier au Professeur Wall). Avant-Propos. *Candollea* **4**, 293–307. 1931.

SAINT-YVES, A. Festucae novae et loci novi Festucarum jam cognitarum (Subgen. Eu-Festuca). *Candollea* **5**, 101–141. 1932.

SOÓ, R. Festuca-Studien. *Acta Bot. Hung.* **2**, 187–221. 1955.

STOHR, G. Der Formenkreis der Festuca ovina L. im mitteldeutschen Trockengebiet. *Wiss. Zeitschr. Univ. Halle Math.-Nat. Reihe* **H4**, 729–746. 1955.

STOHR, G. Gliederung der Festuca ovina-Gruppe in Mitteldeutschland unter Einschluss einiger benachbarter Formen. *Wiss. Zeitschr. Univ. Halle Math.-Nat. Reihe* **9**, 393–414. 1960.

TOWNSEND, F. On the European species of Festuca. *J. Bot.* **20**, 277–281, 302–309. 1882.

LOLIUM

DE ROUVILLE, P. *Monographie du genre Lolium.* Pp. viii + 57. Montpellier. 1853.

ESSAD, S. Contribution à la systematique du genre Lolium. *Ann. Inst. Nat. Rech. Agron. sér. B* **4**, 325–351. 1954.

VULPIA

BLOM, C. Über einige Vulpia-Arten. *Meddel. Göt. Bot. Trad.* **9**, 153–164. 1934.

DUVAL-JOUVE, J. Sur les Vulpia de France. *Rev. Sci. Nat.* **1880**, 18–51. 1880.

HENRARD, J. A study in the genus Vulpia. *Blumea* **2**, 299–326. 1937.

PAUNERO, E. Notas sobre Gramineas, 2. Consideraciones acerca de las especies españolas del género Vulpia Gmel. *Anal. Inst. Bot. Cav.* **22**, 81–114. 1964.

PUCCINELLIA

HOLMBERG, O. R. Einige Puccinellia-Arten und Hybriden. *Bot. Not.* **1920**, 103–111. 1920.

SØRENSEN, T. A revision of the Greenland species of Puccinellia Parl. *Medd. om Grønl.* **136** (3), 1–179. 1953.

POA

ASCHERSON, P. and GRAEBNER, P. *Poa,* in *Synopsis Mitteleur. Fl.* **2** (1), 386–437. 1900.

CHRTEK, J. and VÁCLAV, J. Contribution to the systematics of the genus Poa L. section Ochlopoa (A. et Gr.) V. Jirás. *Preslia* **34**, 40–68. 1962.

JIRÁSEK, V. Einige taxonomische Probleme im Komplex der Poa pratensis L. s.l. *Acta Bot. Hort. Bot. Prag.* **1964**, 60–68. 1964.

LEBAILLY, G. Les espèces du genre Poa L. section Stoloniferae Nannf. indigènes en Belgique. *Bull. Jard. Bot. Bruxelles* **30**, 5–14. 1960.

MARSH, V. L. A taxonomic revision of the genus Poa of the United States and southern Canada. *Amer. Midl. Nat.* **47**, 202–250. 1952.

NANNFELDT, J. Taxonomical and plant geographical studies in the Poa laxa group. *Symb. Bot. Upsal.* **5**, 1–113. 1935.

OETTINGEN, H. von. Kritische Betrachtungen über die Systematik der Gattung Poa L. besonders über die Sektion Pachyneurae Aschers. *Fedde. Rep.* **21**, 306–316, 368. 1925.

DACTYLIS

DOMIN, K. Monografická studie o rodu Dactylis L. *Acta Bot. Bohemica* **14**, 1–147. 1943.

STEBBINS, G. L. and ZOHARY, D. Cytogenetics and evolutionary studies on the genus Dactylis, 1. Morphology, distribution and interrelationship of the diploid subspecies. *Univ. Calif. Publ. Bot.* **31**, 1–40. 1959.

MELICA

PAPP, C. Monographie der europäischen Arten der Gattung Melica L. *Bot. Jahrb.* **65**, 275–348. 1932.

SESLERIA

DEYL, M. Study of the genus Sesleria. *Opera Bot. Čech.* **3**, 1–258. 1946.

UJHELYI, J. Species Sesleriae generis novae. *Fedde. Rep.* **62**, 59–70. 1959.

BROMUS

CAMUS, A. Bromus hybrides de la flora française. *Bull. Jard. Bot. Bruxelles* **28**, 479–485. 1957.

DE CUGNAC, A. Sur quelques Bromes et leurs hybrides. *Bull. Soc. Bot. France* **81**, 318–323. 1934: *loc. cit.* **84**, 437–440, 711–713. 1937: *loc. cit.* **86**, 409–419. 1939.

DE CUGNAC, A. and CAMUS, A. Sur quelques espèces de Bromus et leurs hybrides. *Bull. Soc. Bot. France* **78**, 327–341. 1931: *loc. cit.* **83**, 47–68, 658–667. 1936: *loc. cit.* **88**, 513–517. 1941.

FASSEAUX, W. Le groupe du Bromus mollis L. en Belgique. *Nat. Belge* **32**, 190–196. 1951.

ISELY, D., WEST, D. and POHL, R. W. Seeds of agricultural and weedy Bromus species. *Iowa State Coll. J. Sci.* **25**, 531–548. 1951.

SHEAR, C. L. A revision of the North American species of Bromus occurring north of Mexico. *U.S. Dept. Agric. Agrost. Bull.* **23**, 5–66. 1900.

WAGNON, H. K. A revision of the genus Bromus, section Bromopsis of North America. *Brittonia* **7**, 415–480. 1952.

BRACHYPODIUM

SAINT-YVES, A. Contribution à l'étude des Brachypodium (Europe et Région méditerranéenne). *Candollea* **5**, 427–493 + t.t. xiii–xvii. 1934.

AGROPYRON

HANSEN, A. Elytrigia (Agropyron)-hybrider i Danmark. *Bot. Tidsskr.* **55**, 296–312. 1960.

HANSEN, A. Revision der niederländischen Arten der Gattung Agropyron. *Acta Bot. Neerl.* **10**, 394–396. 1961.

JIRÁSEK, V. Ein Beitrag zur Systematik und Taxonomie der tschechoslowakischen Agropyrum Arten. *Preslia* **26**, 159–176. 1954.

LÖVE, A. Biosystematische Analyse der Elytrigia Junceae Gruppe. *Kulturpflanze Beih.* **3**, 74–85. 1962.

MELDERIS, A. The short-awned species of the genus Roegneria of Scotland, Iceland and Greenland. *Svensk Bot. Tidskr.* **44**, 132–166. 1950.

HORDEUM

BOWDEN, W. M. The taxonomy and nomenclature of the wheats, barleys and ryes and their wild relatives. *Canad. J. Bot.* **37**, 657–684. 1959.

BOWDEN, W. M. Cytotaxonomy of the native and adventive species of Hordeum, Eremopyrum, Secale, Sitarion and Triticum in Canada. *Canad. J. Bot.* **40**, 1675–1711. 1962.

BOWDEN, W. M. Cytotaxonomy of the Eurasiatic and South American species of Barley, genus Hordeum L. *Canad. J. Gen. Cyt.* **7**, 394–399. 1965.

COVAS, G. Taxonomic observations on the North American species of Hordeum. *Madroño* **10**, 1–21. 1949.

NEVSKI, A. Monographiae genus Hordeum. *Acta Acad. Sci. URSS. ser.* 1 **5**, 64–255. 1941.

SECALE

BOWDEN, W. M. The taxonomy and nomenclature of the wheats, barleys and ryes and their wild relatives. *Canad. J. Bot.* **37**, 657–684. 1959.

BOWDEN, W. M. Cytotaxonomy of the native and adventive species of Hordeum, Eremopyrum, Secale, Sitarion and Triticum in Canada. *Canad. J. Bot.* **40**, 1675–1711. 1962.

ROSHEVITZ, R. I. Monographie of the genus Secale L. *Acta Inst. Bot. Acad. Sci. URSS ser.* 1 **6**, 104–163. 1947.

TRITICUM

BOWDEN, W. M. The taxonomy and nomenclature of the wheats, barleys and ryes and their wild relatives. *Canad. J. Bot.* **37**, 657–684. 1959.

BOWDEN, W. M. Cytotaxonomy of the native and adventive species of Hordeum, Eremopyrum, Secale, Sitarion and Triticum in Canada. *Canad. J. Bot.* **40**, 1675–1711. 1962.

DANTHONIA

Vickery, J. W. A revision of the Australian species of Danthonia. *Contr. New South Wales Nat. Herb.* **2**, 249–325. 1956.

KOELERIA

Domin, K. Monographie der Gattung Koeleria. *Bibl. Bot.* **65**, 1–354. 1907.

Ujhelyi, J. Data to the systematics of the subsectio Bulbosae of the genus Koeleria. *Ann. Mus. Hung.* **54**, 199–220. 1962: *loc. cit.* **55**, 187–205. 1963: *loc. cit.* **56**, 195–214. 1964.

Ujhelyi, J. Data to the systematics of the sectio Bulbosae and sectio Caespitosae of the genus Koeleria. *Ann. Mus. Hung.* **57**, 179–202. 1965: *loc. cit.* **58**, 177–196. 1966.

TRISETUM

Chrtek, J. and Jirásek, V. On the taxonomy of the genus Trisetum Pers. *Webbia* **17**, 569–580. 1963.

AVENA

Cosson, M. E. Classification des espèces du genre Avena du groupe de l'Avena sativa (Avena sect. Avenatypus). *Bull. Soc. Bot. France* **1**. Pp. 8. 1854.

Marquand, C. V. B. Varieties of Oats in cultivation. *Rep. Welsh Plant. Breed. Stat.* **6** (2). 1922.

Paunero, E. Las Aveneas españolas, 3. *Anal. Inst. Bot. Cav.* **15**, 377–416. 1957.

Saint-Yves, A. Contribution à l'études des Avena sect. Avenastrum (Eurasie et Région Mediterranéenne). *Candollea* **4**, 353–504. 1930.

Stanton, T. R. Oat identification and classification. *U.S. Dept. Agric. Techn. Bull.* **1100**, 1–206. 1955.

Thellung, A. Über die Abstammung der systematischen Wert und die Kulturgeschichte der Saathaferarten (Avena sativae Cosson). Beiträge zu einer natürlichen Systematik von Avena sect. Eu-Avena. *Viert. Naturf. Ges. Zürich* **56**, 293–350. 1911.

HELICTOTRICHON

Holub, J. *Bemerkungen zur Taxonomie der Gattung Helictotrichon Bess.*, in *Klášterský, I., P. M. Opiz und seine Bedeutung fur die Pflanzentaxonomie*, 101–133. 1958.

ARRHENATHERUM

Kitanov, B. Kritische Bemerkungen über die Vertreter de Gattung Arrhenatherum P.B. auf der Balkanhalbinsel. *Bull. Inst. Bot.* (*Sofia*) **2**, 195–208. 1951.

CORYNEPHORUS

Jirásek, V. and Chrtek, J. Systematische Studie über die Arten der Gattung Corynephorus Pal.-Beauv. (Poaceae). *Preslia* **34**, 374–386. 1962.

CALAMAGROSTIS

Torges, E. Zur Gattung Calamagrostis Adans. *Mitt. Thür. Bot. Ver.* **6**, 14–22. 1894: *loc. cit.* **8**, 13–16. 1895: *loc. cit.* **11**, 78–93. 1897: *loc. cit.* **12**, 22–25. 1898.

Tzvelev, N. De genere Calamagrostis Adans. in URSS notulae systematicae. *Nov. Syst. Plant. Vasc.* **1965**, 5–50. 1965.

Wasilijew, W. N. Das System der Gattung Calamagrostis Roth. *Fedde. Rep.* **63**, 229–251. 1960.

AGROSTIS

Björkman, S. O. Studies in Agrostis and related genera. *Symb. Bot. Upsal.* **17** (1), 1–112. 1960.

Paunero, E. Las especies españolas del género Agrostis. *Anal. Inst. Bot. Cav.* **7**, 561–644. 1947. (1948).

Philipson, W. R. A revision of the British species of Agrostis. *J. Linn. Soc. Bot.* **51**, 73–150. 1937.

PHLEUM

Hall, M. Bibliography on the taxonomy and agricultural botany of herbage Gramineae, 4. Phleum. *Herb. Rev.* **3**, 84–90. 1935.

ALOPECURUS

Paunero, E. Las especies españolas del género Alopecurus. *Anal. Inst. Bot. Cav.* **10**, 301–346. 1952.

ANTHOXANTHUM

Paunero, E. Las especies españolas del género Anthoxanthum. *Anal. Inst. Bot. Cav.* **12**, 401–442. 1954.

PHALARIS

Anderson, D. E. Taxonomy and distribution of the genus Phalaris. *Iowa State Coll. J. Sci.* **36**, 1–96. 1961.

Paunero, E. Revisión de las especies españolas del género Phalaris. *Anal. Inst. Bot. Cav.* **8**, 475–522. 1948.

PARAPHOLIS

Paunero, E. Notas sobre gramineas, 3. Consideraciones acerca de las especies españolas del género Parapholis. *Anal. Inst. Bot. Cav.* **22**, 187–201. 1964.

RUNEMARK, H. A revision of Parapholis and Monerma in the Mediterranean. *Bot. Not.* **115**, 1–17. 1962.

MONERMA

RUNEMARK, H. A revision of Parapholis and Monerma in the Mediterranean. *Bot. Not.* **115**, 1–17. 1962.

SPARTINA

MOBBERLEY, D. G. Taxonomy and distribution of the genus Spartina. *Iowa State Coll. J. Sci.* **30**, 471–574. 1956.

SAINT-YVES, A. Monographia Spartinarum. *Candollea* **5**, 19–100. 1932.

CYNODON

HURCOMBE, R. A cytological and morphological study of cultivated Cynodon species. *J. S. Afr. Bot.* **13**, 107–116. 1947.

ECHINOCHLOA

HITCHCOCK, A. S. The North American species of Echinochloa. *Contr. U.S. Nat. Herb.* **22**, 133–153. 1920.

YABUNO, T. Biosystematic study of the genus Echinochloa. *Jap. J. Bot.* **19**, 277–323. 1966.

DIGITARIA

HENRARD, J. T. *Monograph of the genus Digitaria.* Pp. xxi + 999. Leiden. 1950.

SETARIA

HITCHCOCK, A. S. The North American species of Chaetochloa. *Contrib. U.S. Nat. Herb.* **22**, 155–208. 1920.

HUBBARD, F. T. A taxonomic study of Setaria italica and its immediate allies. *Amer. J. Bot.* **2**, 169–198. 1915.

ROMINGER, J. M. Taxonomy of Setaria (Gramineae) in North America. *Illinois Biol. Monogr.* **29**, 1–132. 1962.

AEGILOPS

EIG, A. Monographisch-Kritische Übersicht der Gattung Aegilops. *Fedde. Rep. Beih.* **55**, 1–228. 1929.

SORGHUM

ROBERTY, G. Monographie systématique des Andropogonées du Globe. *Theses Fac. Sci. Toulouse* **168**, 296–314. 1960.

SNOWDEN, J. D. *The cultivated races of Sorghum.* Pp. vii + 274. London. 1936.

STIPA

CARO, J. A. Las especies de Stipa (Gramineae) de la region central Argentina. *Kurtziana* **3**, 7–119. 1966.

Čelakovsky, L. Über einiger Stipen. *Österr. Bot. Zeitschr.* **33**, 313–344, 349–352. 1883.

Čelakovsky, L. Nachtrag über Stipa Tirsa Steven. *Österr. Bot. Zeitschr.* **34**, 318–321. 1884.

Hitchcock, A. S. The North American species of Stipa. *Contr. U.S. Nat. Herb.* **24**, 215–262. 1925.

Hitchcock, A. S. Synopsis of the South American species of Stipa. *Contr. U.S. Nat. Herb.* **24**, 263–289. 1925.

Martinovský, J. O. Die italienischen "Stipa"-Sippen der sektion "Pennatae". *Webbia* **20**, 711–736. 1965.

ERAGROSTIS

Jirásek, V. Phytogeographical-systematic study of the genus Eragrostis P.B. *Preslia* **24**, 281–338. 1952.

ANDROPOGON

Anderson, J. G. The genus Andropogon in Southern Africa. *Bothalia* **9**, 5–30. 1966.

SASA

Suzuki, S. Taxonomical studies on the Bambusaceous genus Sasa Makino et Shibata. *Jap. J. Bot.* **18**, 289–307. 1964: *loc. cit.* **19**, 99–125. 1965.

Index to Families and Genera, etc.

Synonyms are shown in italics

Abies, 33
Abutilon, 60
Acaena, 71
Acer, 62
Aceraceae, 62
Aceras, 141
Achillea, 121
Acinos, 111
Aconitum, 35
Acorus, 141
Actaea, 37
Adiantaceae, 30
Adiantum, 30
Adonis, 40
Adoxa, 116
Adoxaceae, 116
Aegilops, 152
Aegopodium, 82
Agrimonia, 70
Agropyron, 149
Agrostis, 151
Ajuga, 113
Alchemilla, 70
Alisma, 128
Alismataceae, 127
Alkanna, 100
Alliaria, 49
Allium, 135
Alnus, 89
Alopecurus, 151
Althaea, 60
Alyssum, 46
Amaranthaceae, 56
Amaranthus, 56
Amaryllidaceae, 135
Ambrosia, 119
Amelanchier, 73
Ammi, 82
Amsinckia, 100
Anacamptis, 141
Anagallis, 95
Anchusa, 99
Andropogon, 153
Anemone, 37
Antennaria, 120
Anthoxanthum, 151
Anthyllis, 65
Antirrhinum, 104
Aphanes, 71
Apium, 81
Apocynaceae, 96
Aponogeton, 131
Aponogetonaceae, 131
Aquifoliaceae, 62
Aquilegia, 40
Arabidopsis, 49
Arabis, 48
Araceae, 141
Araliaceae, 80
Arctium, 123
Arctostaphylos, 92
Arctous, 92
Arenaria, 55
Argemone, 42
Aristolochia, 84
Aristolochiaceae, 84
Armeria, 94
Arrhenatherum, 150
Artemisia, 122
Arum, 141
Arundineae, 145
Asarina, 104
Asarum, 84
Asparagus, 133

Asperugo, 98
Asperula, 115
Aspidiaceae, 31
Aspleniaceae, 30
Asplenium, 30
Aster, 121
Astragalus, 66
Astrantia, 81
Athyriaceae, 31
Athyrium, 31
Atriplex, 58
Atropa, 102
Aubrieta, 48
Avena, 150
Axyris, 57
Azolla, 32
Azollaceae, 32

Baldellia, 128
Ballota, 112
Barbarea, 47
Berberidaceae, 41
Berberis, 41
Berula, 82
Beta, 57
Betonica, 112
Betula, 88
Betulaceae, 88
Bidens, 118
Biscutella, 45
Blackstonia, 97
Boraginaceae, 98
Borago, 99
Botrychium, 32
Brachypodium, 149
Brassica, 44
Bromus, 148
Bryonia, 83
Bunium, 82
Bupleurum, 81
Butomaceae, 129
Butomus, 129
Buxaceae, 62
Buxus, 62

Cakile, 45
Calamagrostis, 151
Calamintha, 111
Calendula, 118
Calla, 141
Callitrichaceae, 79
Callitriche, 79
Calluna, 92
Calotis, 122
Caltha, 35
Calystegia, 101
Camelina, 49
Campanula, 114
Campanulaceae, 114
Caprifoliaceae, 116
Capsella, 46
Capsicum, 103
Cardamine, 47
Cardaria, 46
Carduus, 123
Carex, 144
Carpinus, 89
Carthamus, 124
Carum, 82
Caryophyllaceae, 52
Castanea, 89
Celastraceae, 62
Centaurea, 124
Centaurium, 97
Cephalaria, 118
Cerastium, 53
Ceratophyllaceae, 42
Ceratophyllum, 42
Chaenorhinum, 104
Chamaecyparis, 34
Chamaenerion, 77
Chelidonium, 43
Chenopodiaceae, 57
Chenopodium, 57
Cherleria, 54
Chrysanthemum, 122
Chrysosplenium, 76
Cicendia, 97
Cicer, 67
Cicerbita, 125
Cicuta, 82
Circaea, 78
Cirsium, 123
Cistaceae, 51

Cistus, 51
Citrullus, 83
Cladium, 144
Clematis, 37
Clinopodium, 111
Cochlearia, 46
Coeloglossum, 139
Colchicum, 133
Colutea, 66
Compositae, 118
Conringia, 45
Consolida, 36
Convolvulaceae, 101
Convolvulus, 101
Conyza, 121
Cornaceae, 80
Cornus, 80
Coronilla, 66
Coronopus, 46
Corydalis, 43
Corylaceae, 89
Corylus, 89
Corynephorus, 151
Cosmos, 120
Cotoneaster, 73
Cotula, 122
Crambe, 45
Crassulaceae, 74
Crataegus, 73
Crepis, 126
Crocus, 137
Crucianella, 116
Cruciata, 116
Cruciferae, 43
Cucumis, 83
Cucurbita, 83
Cucurbitaceae, 83
Cupressaceae, 34
Cuscuta, 101
Cyclamen, 95
Cymbalaria, 104
Cynodon, 152
Cynoglossum, 98
Cyperaceae, 148
Cyperus, 144
Cypripediaceae, 138
Cypripedium, 138
Cystopteris, 31
Cytisus, 63

Dactylis, 148
Dactylorchis, 140
Dactylorhiza, 140
Damasonium, 129
Danthonia, 150
Daphne, 77
Datura, 103
Daucus, 83
Delphinium, 36
Dennstaedtiaceae, 30
Descurainia, 49
Dianthus, 52
Diapensia, 93
Diapensiaceae, 93
Dicentra, 43
Dichondra, 101
Digitalis, 105
Digitaria, 152
Diphasium, 27
Diplotaxis, 44
Dipsacaceae, 117
Dipsacus, 117
Doronicum, 120
Dorycnium, 66
Draba, 47
Drosera, 76
Droseraceae, 76
Dryas, 70
Dryopteris, 31
Duchesnea, 69

Echinochloa, 152
Echinops, 123
Echium, 100
Egeria, 130
Elatinaceae, 52
Elatine, 52
Eleagnaceae, 77
Eleocharis, 143
Elodea, 130
Empetraceae, 94
Empetrum, 94
Epilobium, 77
Epimedium, 41

Epipactis, 139
Epipogium, 139
Equisetaceae, 28
Equisetum, 28
Eragrostis, 153
Eranthis, 35
Erica, 92
Ericaceae, 91
Erigeron, 121
Eriocaulaceae, 132
Eriocaulon, 132
Eriophorum, 143
Erodium, 61
Erophila, 47
Eruca, 45
Erucastrum, 44
Eryngium, 81
Erysimum, 48
Eschscholzia, 43
Euonymus, 62
Euphorbia, 84
Euphorbiaceae, 84
Euphrasia, 107

Fagaceae, 89
Fagopyrum, 85
Falcaria, 82
Festuca, 146
Ficus, 88
Filago, 121
Filipendula, 68
Fragaria, 70
Fraxinus, 96
Fritillaria, 133
Fuchsia, 78
Fumaria, 43
Fumariaceae, 43

Gagea, 133
Galanthus, 136
Galeobdolon, 112
Galeopsis, 112
Galinsoga, 119
Galium, 115
Genista, 63
Gentiana, 97
Gentianaceae, 97
Gentianella, 97
Geraniaceae, 61
Geranium, 61
Geum, 70
Gladiolus, 138
Glaucium, 42
Glaux, 96
Glechoma, 113
Glyceria, 145
Glycine, 68
Gramineae, 145
Grossulariaceae, 76
Gunnera, 79
Guttiferae, 50
Gymnadenia, 139
Gymnospermae, 33
Gypsophila, 53

Halimione, 58
Haloragaceae, 68
Hedera, 80
Helianthemum, 51
Helianthus, 119
Helictotrichon, 150
Helleborus, 35
Heracleum, 83
Herminium, 139
Hermodactylus, 137
Herniaria, 55
Hesperis, 48
Hibiscus, 60
Hieracium, 125
Himantoglossum, 140
Hippocrepis, 67
Hippophae, 77
Hirschfeldia, 44
Holodiscus, 68
Holosteum, 54
Honkenya, 55
Hordeum, 149
Huperzia, 26
Hydrocharis, 130
Hydrocharitaceae, 130
Hydrocotyle, 80
Hymenophyllaceae, 29
Hymenophyllum, 30
Hypericum, 51

Hypochoeris, 124
Hyssopus, 110

Ilex, 62
Illecebraceae, 55
Inula, 120
Ipomoea, 101
Iridaceae, 137
Iris, 137
Isoetaceae, 28
Isoetes, 28
Iva, 123

Juglandaceae, 88
Juglans, 88
Juncaceae, 134
Juncaginaceae, 131
Juncus, 134
Juniperus, 34

Knautia, 117
Kochia, 58
Koeleria, 150
Kohlrauschia, 53

Labiatae, 108
Lactuca, 125
Lagarosiphon, 131
Lagenaria, 83
Lamiastrum, 112
Lamium, 112
Lappula, 100
Lapsana, 124
Larix, 33
Lathraea, 107
Lathyrus, 67
Lavatera, 59
Ledum, 91
Legousia, 115
Leguminosae, 63
Lemna, 142
Lemnaceae, 142
Lentibulariaceae, 108
Leontodon, 124
Leonurus, 112
Lepidium, 45
Lepidotis, 26
Leucojum, 136
Leucorchis, 140
Ligustrum, 96
Liliaceae, 132
Lilium, 133
Limonium, 94
Limosella, 105
Linaceae, 60
Linaria, 104
Linnaea, 116
Linum, 60
Liparis, 139
Lithospermum, 100
Littorella, 114
Lobelia, 115
Lolium, 147
Lonicera, 116
Loranthaceae, 79
Lotus, 66
Lupinus, 63
Luronium, 128
Luzula, 135
Lychnis, 52
Lycium, 102
Lycopodiaceae, 25
Lycopodiella, 26
Lycopodium, 26
Lycopus, 109
Lysichiton, 141
Lysimachia, 95
Lythraceae, 76
Lythrum, 76

Mahonia, 41
Maianthemum, 133
Malcolmia, 48
Malus, 74
Malva, 59
Malvaceae, 59
Marrubium, 113
Matthiola, 48
Meconopsis, 42
Medicago, 64
Melampyrum, 106
Melica, 148
Melilotus, 64
Melissa, 111

Melittis, 111
Mentha, 109
Menyanthaceae, 98
Mercurialis, 84
Mertensia, 100
Mimulus, 104
Minuartia, 54
Moehringia, 55
Molinia, 145
Monerma, 152
Moneses, 93
Montia, 56
Moraceae, 88
Myosotis, 99
Myosurus, 40
Myrica, 88
Myricaceae, 88
Myriophyllum, 78

Najadaceae, 132
Najas, 132
Narcissus, 136
Nasturtium, 47
Neotinea, 140
Nepeta, 112
Nicotiana, 102
Nigella, 36
Nuphar, 42
Nymphaea, 41
Nymphaeaceae, 41
Nymphoides, 98

Odontites, 107
Oenanthe, 82
Oenothera, 77
Oleaceae, 96
Omphalodes, 98
Onagraceae, 77
Onobrychis, 67
Ononis, 64
Onopordum, 124
Ophioglossaceae, 32
Ophioglossum, 33
Ophrys, 140
Orchidaceae, 138
Orchis, 140
Origanum, 109
Ornithogalum, 133
Orobanchaceae, 107
Orobanche, 107
Orthilia, 93
Osmunda, 29
Osmundaceae, 29
Oxalidaceae, 61
Oxalis, 61
Oxytropis, 66

Paeonia, 40
Paeoniaceae, 40
Papaver, 42
Papaveraceae, 42
Parapholis, 151
Parietaria, 87
Paris, 134
Paronychia, 56
Parthenocissus, 63
Pedicularis, 106
Pentaglottis, 99
Peplis, 76
Petrorhagia, 53
Petroselinum, 81
Peucedanum, 83
Phalaris, 151
Phaseolus, 68
Phleum, 151
Phlomis, 112
Phyllodoce, 92
Physalis, 102
Phyteuma, 115
Phytolacca, 59
Phytolaccaceae, 59
Pilosella, 126
Pimpinella, 82
Pinaceae, 33
Pinguicula, 108
Pinus, 33
Plantaginaceae, 113
Plantago, 113
Platanaceae, 88
Platanthera, 140
Platanus, 88
Plumbaginaceae, 94
Poa, 147
Polemoniaceae, 98

Polemonium, 98
Polygala, 50
Polygalaceae, 50
Polygonaceae, 84
Polygonatum, 132
Polygonum, 85
Polypodiaceae, 32
Polypodium, 32
Polystichum, 32
Populus, 90
Portulaca, 56
Portulacaceae, 56
Potamogeton, 131
Potamogetonaceae, 121
Potentilla, 69
Poterium, 71
Primula, 95
Primulaceae, 95
Prunella, 111
Prunus, 72
Pseudorchis, 140
Psoralea, 64
Pteridium, 30
Pteridophyta, 25
Pteris, 30
Pteropsida, 29
Puccinellia, 147
Pulmonaria, 99
Pulsatilla, 37
Pyrola, 93
Pyrolaceae, 92

Quercus, 90

Ranunculaceae, 34
Ranunculus, 37
Raphanus, 45
Rapistrum, 45
Reseda, 49
Resedaceae, 49
Rhamnaceae, 62
Rhamnus, 62
Rhinanthus, 106
Rhododendron, 92
Rhynchosinapis, 44
Rhynchospora, 144
Ribes, 70
Roemeria, 42
Romulea, 137
Rorippa, 48
Rosa, 72
Rosaceae, 68
Rubiaceae, 115
Rubus, 68
Rumex, 85
Ruppia, 132
Ruppiaceae, 132

Sagina, 54
Sagittaria, 129
Salicaceae, 90
Salicornia, 58
Salix, 90
Salsola, 58
Salvia, 111
Sambucus, 116
Samolus, 96
Sanguisorba, 71
Sanicula, 80
Santalaceae, 79
Saponaria, 53
Sarothamnus, 63
Sasa, 153
Satureja, 110
Saxifraga, 75
Saxifragaceae, 75
Scabiosa, 118
Scheuchzeria, 131
Scheuchzeriaceae, 131
Schkuhria, 119
Schoenus, 144
Scirpus, 143
Scleranthus, 56
Scorpiurus, 67
Scorzonera, 125
Scrophularia, 104
Scrophulariaceae, 103
Scutellaria, 113
Secale, 149
Sedum, 74
Selaginella, 27
Selaginellaceae, 27
Sempervivum, 75
Senecio, 120

Sesleria, 148
Setaria, 152
Sibbaldia, 69
Sibthorpia, 105
Sida, 60
Sidalcea, 60
Sigesbeckia, 118
Silene, 52
Sinapis, 44
Sison, 81
Sisymbrium, 49
Sisyrinchium, 137
Sium, 82
Solanaceae, 102
Solanum, 102
Solidago, 121
Sonchus, 125
Sorbus, 74
Sorghum, 152
Sparganiaceae, 142
Sparganium 142
Spartina, 152,
Spartium, 63
Spergularia, 55
Spinacia, 58
Spiraea, 68
Spirodela, 142
Stachys, 111
Stellaria, 54
Sternbergia, 136
Stipa, 152
Stratiotes, 130
Suaeda, 58
Subularia, 46
Succisa, 118
Swida, 80
Symphoricarpos, 116
Symphytum, 98

Tamaricaceae, 51
Tamarix, 51
Taraxacum, 126
Taxaceae, 34
Taxus, 34
Tetragonolobus, 66
Teucrium, 113
Thalictrum, 40
Thelycrania, 80
Thelypteridaceae, 32
Thelypteris, 32
Thesium, 79
Thlaspi, 46
Thymelaceae, 77
Thymus, 109
Tilia, 59
Tiliaceae, 59
Tofeldia, 132
Tragopogon, 125
Trichomanes, 29
Trientalis, 95
Trifolium, 64
Triglochin, 131
Trigonella, 65
Trinia, 81
Trisetum, 150
Triticum, 149
Trollius, 35
Tropaeolaceae, 60
Tropaeolum, 60
Tuberaria, 51
Tulipa, 133
Turritis, 48
Typha, 143
Typhaceae, 143

Ulex, 63
Ulmaceae, 87
Ulmus, 87
Umbelliferae, 80
Umbilicus, 75
Urtica, 87
Urticaceae, 86
Utricularia, 108

Vaccinium, 92
Valeriana, 117
Valerianaceae, 116
Valerianella, 116
Vallisneria, 131
Verbascum, 103
Verbena, 108
Verbenaceae, 108
Verbesina, 119
Veronica, 105

Viburnum, 116
Vicia, 67
Vinca, 96
Viola, 49
Violaceae, 49
Viscum, 79
Vitaceae, 62
Vitis, 63
Vulpia, 147

Wahlenbergia, 114
Wissadula, 60
Wolffia, 142
Woodsia, 31

Xanthium, 119

Zostera, 131
Zosteraceae, 131